Okamura · Rånby · Ito (Eds.)

Macromolecular Concept and Strategy for Humanity
in Science, Technology and Industry

Springer
Berlin
Heidelberg
New York
Barcelona
Budapest
Hong Kong
London
Milan
Paris
Santa Clara
Singapore
Tokyo

Okamura · Rånby · Ito (Eds.)

Macromolecular Concept and Strategy for Humanity in Science, Technology and Industry

With contributions by
Y. Ito, S. Kawabata, M. Niwa, S. Okamura, B. Rånby,
Y. Sakurada, K. Takakura, N. Yoda

With 67 Figures and 51 Tables

 Springer

Dr. Y. Ito
Toray Industries, Inc., 2-1-1 Nihonbashi Muromachi, Chuo-ku Tokyo
103 Japan

Prof. Dr. S. Okamura
Kyoto University, 24 Minamigoshomachi, Okazaki, Sakyo-ku Kyoto
606 Japan

Prof. Dr. B. Rånby
Royal Inst. of Technology, Dept. of Polymer Technology, Stevenbocksvagen 21
18262 Djurshölm, Sveden

ISBN-13:978-3-642-64665-2 Springer-Verlag Berlin Heidelberg New York

Cataloging-in-Publication Data applied for

Die Deutsche Bibliothek – CIP-Einheitsaufnahme
**Macromolecular concept and strategy for humanity in science,
technology and industry** / Okamura ; Rånby ; Ito (Hrsg.). With
contributions by Y. Ito ... – Berlin ; Heidelberg ; New York ;
Barcelona ; Budapest ; Hong Kong ; London ; Milan ; Paris ;
Santa Clara ; Singapore ; Tokyo : Springer, 1996
    ISBN-13:978-3-642-64665-2   e-ISBN-13:978-3-642-61036-3
    DOI: 10.1007/978-3-642-61036-3

NE: Okamura, Seizo [Hrsg.]; Ito, Yoshikazu

Production: PRODUserv Springer Produktions-Gesellschaft, Berlin
Typesetting: Graphische Werkstätten Lehne GmbH, Grevenbroich
SPIN: 10499552   2/3020-5 4 3 2 1 0 – Printed on acid-free paper

Preface

This book is to reconsider the field of polymer chemistry in the eyes of both naturalists and historians, in an attempt to perceive the evolution of its concepts and strategies as an array of trial and error. Namely, in retrospect, the origin of polymer chemistry may date back to the beginning of this century, where the then colloid chemistry dealt with the size of particles through understanding the *state* of substances including what we now coin polymers. At that time, however, little was known about the *essence* of substances, resulting in little progress in clarifying the particle sizes and therefore the nature of polymers. It was exactly under such circumstances where polymer chemistry began to evolve.

In the following half a century (1920-1970), the spectacular advance in polymer chemistry extensively revealed the *essence* of polymers, in one hand, by the synthesis of polymers (individuals) coupled with the development of synthetic polymer materials (their assembly) and, in the other, by the combination of mechanistic approach (or synthetic polymer chemistry) and biological approaches (or natural polymer chemistry). It is therefore expected that, in the coming half a century (1970-2020), the *state* of polymers, whose *essence* has now been established, will be fully studied and clarified. Such advance in polymer chemistry may also contribute in completing the original object of the early colloid chemistry to understand the size and state of particles in general.

As its title implies, this book, *Macromolecular Concept and Strategy for Humanity in Science, Technology and Industry*, is a compilation of articles where evolution and advances in macromolecular concepts and strategies are discussed in relation to humanity, not purely chronologically, not purely scientifically, but in some combination of both ways blended with interest in humanity. Behind this new approach lies the Editors' belief that humanity has played, and should play, a crucial role in evolution and development of polymer chemistry, as symbolized by my first wording in this preface, „an array of try and error". Perhaps it is for the first time that the field of polymer chemistry is systematically treated with emphasis in humanity, and in this regard we believe that the book will be able to shed new light on the history and the current status of polymer chemistry and polymer industry. The organization of the book, with 11 Chapters in three Parts, is described in Chapter 1.

As one of the Editors of the book, I express my sincere thanks to all of the authors, who contributed interesting chapter articles, to Professor Bengt Rånby and Dr. Naoya Yoda for their editorial assistance, and to Dr. Marion Hertel and Ms. Andrea Weber of Springer for their help in publishing the book.

September 1995   *Seizo Okamura*

Contributing Authors

Dr. Y. Ito
Toray Industries, Inc., 2–1–1 Nihonbashi Muromachi, Chuo-ku Tokyo
103 Japan

Prof. Dr. S. Kawabata
Kyoto University, Dept. of Polymer Chemistry, Kyoto
606 Japan
Current address: Department of Material Science, The University of Shiga Prefecture,
2500 Hassaka, Hikone
522 Japan

M. Niwa
Department of Clothing Science, Nara Women's University, Nara
630 Japan

Prof. Dr. S. Okamura
Kyoto University, 24 Minamigoshomachi, Okazaki, Sakyo-Ku Kyoto
606 Japan

Prof. Dr. B. Rånby
Royal Inst. of Technology, Dept. of Polymer Technology, Stevenbocksvagen 21
18262 Djurshölm, Sweden

Dr. Y. Sakurada
Haemonetics Japan Co., Ltd., Chiyoda-Ku Tokyo
102 Japan

Dr. K. Takakura
Kuraray Co., Ltd., Medical Products Div., 1–12–39 Umeda, Kita-Ku Osaka
530 Japan

Prof. Dr. N. Yoda
Toray Corporate Business Research, Inc., 1–8–1 Mihama, Urayasu-City
279 Japan
Current address: Keio University, Faculty of Science and Technology, 6–45–24 Hino-
minami, Konan-Ku, Yokohama
234 Japan

Contents

Chapter 1

Introduction

SEIZO OKAMURA

This book discusses the macromolecular concept in academia and the macromolecular strategy in industry, focusing on the humanity in society in relation to the science, technology, and industry of polymers and polymeric materials. The seven coauthors, four from academia (Okamura, Rånby, Kawabata, and Takakura) and three from industry (Ito, Yoda, and Sakurada) in accordance with the above-mentioned scope of the book, are very experienced scientists and entrepreneurs who have been actively engaged in polymer research.

This book consists of three parts. Part One concerns the history of macromolecules and the current trend in polymer science; Part Two focuses on the applications of polymers to plastics, rubber, textile, and medicine; Part Three deals with the strategies in chemical industry in general and polymer industry in particular.

In Part One, Chaps. 2 and 3 (both by B. Rånby, Stockholm Technical University) discuss the history of macromolecular concept and science that may consist of three stages: the first was the production and modification of natural and biological products, which was, however, not based on any exact knowledge of macromolecules per se in terms of their concept, structure, and properties; the second stage involved the emergence and the subsequent acceptance of the concept of macromolecules, followed by the third stage where polymer science and technology have apparently been reaching maturity. These two chapters are thus intended to present readers with a long-range perspective of polymer science from evolution to maturity.

It will also be emphasized therein that, reaching some stage of maturity, macromolecular science and concepts now increasingly exert a strong impact on the scientific community and chemical industry, as witnessed for completely new advanced polymeric materials that significantly contribute to today's high technology. The editors and the coauthors of this book, therefore, believe that, in the coming years and perhaps in the next century, the macromolecular concept should be applied, with deep humanity, for the benefit of society.

To conclude Part One, Chap. 4 (S. Okamura, Kyoto University) presents an account of new trends in state-of-the-art science and technology and also of the relationships between technology and arts. In the author's view, bringing technology and arts together is one way to bridge the gap between society and the technological community. In this chapter, more specifically, the society-technology relationships through polymeric materials is discussed by focusing on their properties, functions, and sensitivity, particularly in the context of bio-medical compatibility and functions.

Okamura, Rånby, Ito (Eds.): Macromolecular Concept and Strategy
© Springer-Verlag Berlin · Heidelberg 1996

In Part Two, Chap. 5 (B. Rånby) concerns plastics and rubber, Chap. 6 (N. Yoda, Toray Industries, Inc.) and Chap. 7 (S. Kawabata, Kyoto University) are directed toward textiles, and Chap. 8 (Y. Sakurada, Haemonetics Japan Company and Kyoto University and K. Takakura, Kuraray Company and Okayama University of Science) deals with medical applications of polymers. Thus, Chaps. 5–7 give new viewpoints on plastics, rubber, and textile applications on the basis of macromolecular concepts. Chapter 8 emphasizes that the medical applications of polymers, as one of the most noticeable technologies to date, seem to be among the best ways to achieve a good "public acceptance" (PA) of advanced technologies in society as some kind of art.

In Part Three, Chap. 9 (N. Yoda) and Chap. 10 (Y. Ito) focus on the "cooperation and globalization problems" in the materials industry and chemical industrys respectively. The coauthors of the two chapters both come from Toray Industries, Inc. where Ito is the chairman and Yoda is the president of Toray Corporate Business Research, Inc. Dr. Ito also formerly served as President of the Chemical Society of Japan and as chairman of the Federation & Management of Chemical Industry, Japan (Kobunshi Doyokai). In Chap. 11, Okamura first describes the reason why this book is published as a volume of the *Advances in Polymer Science* series and then proceeds to discuss another important problem, the connection of polymer science and chemical technologies with human body and mind, which is a typical example of the fusion phenomena of different cultures and disciplines that have also been treated in Chaps. 9 and 10.

For the problem of so-called "public acceptance" (PA), we have had, and still have, many attempts to achieve intimate connections between society and the technological community. In these attempts, it should be mandatory to find the same or at least similar opinions on both sides. At present, unfortunately, several leading technologies are still beyond full understanding of the general public; toward these technologies some people are strongly affirmative while some are extremely negative.

Although society and the current advanced technology certainly have different goals, it is equally true that, in order to bridge a gap between them, both communities should share some common views and opinions. The society-technology relationship may be similar to that between arts and science, as clearly expressed by Professor T. S. Kuhn in his second book: : "Science has similar processes but different goals with arts." Such a perception that connects science with arts may also be beneficial when contemplating the relation between science/technology and humanity, which is the subject that I wished to discuss when first planning to edit this volume.

Chapter 2

The Concept of Macromolecules-Emergence, Development and Acceptance

Bengt Rånby

2.1
Introduction

Macromolecular materials are now an essential part of every-day life. We can hardly imagine our existence without plastics, synthetic rubber, fibers, paints and glues. Young people may take these materials for granted. Yet the common use of synthetic macromolecular materials dates back only about 50 years.

In his remarkable book, *Polymers. The Origins and Growth of a Science,* published in 1985, Herbert Morawetz has seen the development of polymers as materials in three periods [2.1].

During *the early period*, until the outbreak of the first world war in 1914, native (biological) macromolecules were used and even modified without basic knowledge of the chemical (molecular) structure of the substances. Natural rubber, starch and cellulose were considered to be colloidal materials without much discussion of their true molecular structure. The industrial development of modified materials was rather impressive, e.g. vulcanization of rubber for tires and shoe soles, production of dextrin from starch, spinning of cellulose fibers from viscose and synthesis of cellulose nitrate for gunpowder, celluloid and fibers. Even the first synthetic material, phenol-formaldehyde resin, was invented and marketed as Bakelite during this early period (1907–09 by Leo Baekeland).

The second period – from 1914 to 1942 – is regarded by Morawetz to be the classical era of polymer science. This period saw the emergence and acceptance of the macro-molecular concept with Hermann Staudinger as the champion. The period was domi-nated by the pioneers of polymer science, Herman F. Mark, Kurt H. Meyer, Werner Kuhn, Wallace H. Carothers and the early work of Paul J. Flory. They established polymers as long-chain molecules which were flexible coils in polymer solutions.

They worked out the kinetics of polymerization in detail, interpreted the X-ray diffraction patterns of crystallized polymers like cellulose, polyamides (nylon) and linear polyesters, and also derived a thermodynamic theory for rubber elasticity based on the flexible polymer chain model.

The third period – from 1942 to 1960 – is regarded by Morawetz as the time when polymer science reached full maturity. The statistical mechanics of flexible and stiff chains was worked out, new polymer structures were synthesized, e.g. polycarbonates,

Okamura, Rånby, Ito (Eds.): Macromolecular Concept and Strategy
© Springer-Verlag Berlin · Heidelberg 1996

aromatic polyamides, polyimides, linear polyethylene and stereoregular polyolefins and polydienes. The spectacular growth of the polymer industry was initiated during this period. Paul J. Flory published his book *Principles of Polymer Chemistry* in 1953 and it is still a valid and authorative textbook on basic polymer science [2.2]

For practical reasons Morawetz's story of polymer science is cut off at 1960. The rapid development has continued during the recent decades, e.g., in the successful synthesis and study of many *liquid crystalline polymers* (LCP) and the *electrical conducting polymers* (ECP), the latter prepared by "doping" of conjugate polymer structures. In the ECP materials, the polymer chains are carriers of unpaired electrons and delocalized electron pairs which are mobile and give electrical, magnetic and optical properties of great interest in solid state physics as well as in their growing applications. The discovery and development of ECP materials is a milestone in polymer science deseribed at a recent Nobel symposium.

In the following discussion it is important to observe that a sizeable industrial development and production of polymeric materials occurred *before* the macromolecular concept was formulated and even considered. The impact of the macromolecular concept – once it had been accepted – on polymer science and biochemistry, biophysics and medicine has been and still is tremendous. Polymer science is now a main driving force in the industrial development of new materials for advanced technologies, e.g. in aeroplanes, automobiles, electronics, computers, telecommunication and medical technology. New constructions and new systems for new applications are made possible by the new advanced materials.

The approach in this chapter is to show how polymer science and technology has its origin in the early practical use of natural polymeric materials. The introduction and acceptance of the macromolecular concept caused a shift in emphasis from empirical knowledge to scientific research and from natural to synthetic polymers in the development of polymeric materials on a large scale.

2.2
The Early Polymer Industry

Man has used polymeric materials in the form of wood, skin and fibers since prehistoric times. Man gradually invented techniques to modify these natural materials, e.g. to make leather from hides, to dye fibers for making fabric and to prepare paints and glues.

2.2.1
Rubber Industry

The early use of natural rubber from Hevea trees in South and Middle America is a fascinating story [2.3]. Hevea rubber was known by the American Indians when Columbus arrived to Haiti in 1492. The Indians played with bouncing balls from the coagulated sap of the wild trees which they called "cahutchu", from which the French word "caoutchouc" is derived. This material was brought to Europe in the 1700's where Joseph Priestly in England called it "rubber" (1770) because of its ability to remove the marks of black pencil from paper by rubbing. Another invention was made in 1823 by Charles Macintosh in England who bonded two layers of fabric with Hevea rubber to produce a waterproof sheet for raincoat, which still is called a "macintosh".

The early use of Hevea rubber was hampered by its deficient properties. It was sticky above room temperature and turned brittle in the cold. After many attempts to remedy these shortcomings by exposure to sunlight, mixing with flour etc., the vulcanization process by heating with sulfur was invented by Charles Goodyear in the USA (1839). The invention was accidental and Goodyear spent several years developing his process. His patent application was eventually granted as a US patent on June 15, 1844. In the meantime Thomas Hancock in England had applied for a patent on vulcanization of rubber with sulfur which was awarded as a British patent on May 21, 1844. Charles Goodyear never accepted Thomas Hancock as the inventor of vulcanization.

The vulcanization process initiated a rapid growth of the rubber industry. After the invention of the pneumatic tire by John Dunlop in England in 1888, the automobile industry became the largest market for rubber, and it still is. The rubber production had grown to 6000 tons in 1860 and grew further to 50 000 tons in 1900. This industrial production and processing was made without basic knowledge of the polymer chain structure of the rubber or the chemical nature of the vulcanization process.

2.2.2
Cellulose-Based Industry

Wood chemistry has been studied for about 150 years and industrial processes for isolation of cellulose fibers from wood have been known for more than a hundred years. Cellulose was first identified and described by Payen [2.4] in 1839 using hydrolysis of wood and other plant materials with hot dilute nitric acid to remove "incrusting material" later called "lignin". Payen found the same fibrous material with the composition 44 % C and 56 % H_2O – which is correct for $C_5H_{10}O_5$ – in various plant cell walls and named it "cellulose". The incrusting material had higher C content (about 65 %). Technical methods to prepare paper fibers from wood were invented in the 1870's, i.e. the acid sulphite and alkaline sulphate processes, without knowledge of the macromolecular structure of the cellulose and the lignin involved [2.5].

A cellulose-based industry was started more than a hundred years ago. The nitration of cotton fibers was first described in 1833. The plasticizing of cellulose nitrate with camphor to celluloid was invented in 1865 [2.6]. This material, although highly inflammable, was manufactured and used for combs, doll heads, stiff collars and cuffs, billiard balls, toothbrushes and even dentures in fairly large quantities in the 1880s. The use of celluloid as film base for the emerging motion picture industry started around 1900. Because of serious fire accidents due to its flammability, celluloid was soon replaced by cellulose acetate as film base. Another important invention was the viscose process by Cross and Bevan in England in 1892. Native cellulose was converted to a water soluble derivative, cellulose xanthate, by treatment with caustic soda and carbon disulfide.

Cellulose fiber ("viscose rayon") and cellulose film ("cellophane") were produced by extrusion of the aqueous solution of cellulose xanthate ("viscose") into an aqueous sulfuric acid-sodium sulfate solution and washing with water. This meant a regeneration of the cellulose material.

These solution and regeneration processes were developed and used in large fiber and film plants in Western Europe and USA in the early 1900's.

The production of rayon and cellophane was accomplished without knowledge or much discussion in the industry about the molecular structure of the cellulose and the reactions involved in the various processes. Among scientists the structure and chemical reactions of cellulose and lignin were studied without much influence on the industrial developments.

2.2.3
Phenol-Formaldehyde Resin (Bakelite)

The chemical reaction between phenol and formaldehyde had already been studied in Germany in the 1870s and the formation of a spongy brittle phenol-formaldehyde material observed in the 1880s.

The practical use of phenol-formaldehyde products as a plastic material (Bakelite) reinforced with sawdust or cellulose fibers was invented by Leo H. Baekeland in 1907. A US patent was granted and production of Bakelite was started in USA and Germany in 1910 [2-7]. This was the first synthetic polymeric material produced commercially and it is still, 80 years later, an important material for industrial applications, due to its mechanical strenght and durability. Again we find that the synthesis and the invention of reinforcements and the successful production and use of a polymeric material was made without knowledge of the macromolecular structure of the product.

2.3
Early Studies of the Molecular Structure of Native Substances

2.3.1
Hevea Rubber

Due to its unique mechanical properties, natural rubber from Hevea trees was an exciting material to scientists in the 1700s and 1800s. Michael Faraday was among the first to make a careful chemical analysis of Hevea rubber [2.8]

He found a C/H weight ratio of 6. 812 (correct value for C_5H_8 is 7. 50) and reported that Hevea rubber resembled albumin (the protein in eggwhite). Faraday's rubber samples were probably contaminated with proteins of the Hevea latex from which the rubber had been precipitated.

The studies of Hevea rubber continued by pyrolysis which yielded an organic vapor which was eventually correctly identified as isoprene (C_5H_8) by German chemists about 1900. How isoprene materials constituted rubber was a very difficult research problem. The most influential work was by Harries in the early 1900s who argued that rubber consists of cyclic dimers of isoprene, held together by attraction of the double bonds [2.9] The first scientist to propose the correct poly(isoprene) chain structure for Hevea rubber was Pickles [2.10]. His problem was the end groups which he eventually eliminated by assuming that the poly(isoprene) chains formed large ring structures.

The elastic properties of rubber fascinated scientists and led to thermodynamic measurements and much speculation. The first correct observations of the nature of rubber elasticity were made by Gough in 1805 [2.11] using a rubber band and his lips to indicate temperature changes – on sudden stretching warming, and on subsequent contraction cooling. These experiments and the observation that a stretched rubber band contracted on heating could not be correctly interpreted during the 1800s due to the lack of an adequate molecular model. Not before 1934, when the macromolecular concept was established and accepted, could Eugene Guth and Herman Mark initiate the development of a kinetic theory of rubber elasticity based on the assumption of flexible polymer chains with free rotation around the bonds. They calculated the retraction force when the end-to-end distance of a single chain was extended [2.12] This approach has been successfully used in later work on the theory of rubber elasticity by Kuhn, Anthony and Flory. The theoretical and experimental studies of rubber elasticity has been important for the understanding of polymer chain conformation, dynamics of chains and the deformation of polymer chains in solution, melt and solid state.

2.3.2
Starch

The chemical structure of starch and cellulose was a central problem in chemical research all through the 1800s. That starch is converted to sugar by acid and that both sugar and starch contain oxygen and hydrogen in the same proportions as water became known in 1811 [2.13]. The name "carbohydrate" is based on this analysis.

The observation made in 1814 that starch is colored blue by mixing with iodine has been important for analysis in carbohydrate research [2.13]. The optical activity of starch products in solution is another property for identification which gave hydrolyzed starch the name "dextrin" due to its strong rotatory power.

That enzymatic hydrolysis of starch gives a disaccharide, maltose, and not glucose as previously assumed, was reported in 1876 [2.14]. Molecular weight determination was a difficult problem due to the lack of adequate methods. In the 1890s, cryoscopic measurements gave molecular weights of 20 000 to 30 000 for starch and about one fifth for dextrin [2.15]. End group analysis of slightly oxidized starch and dextrin supported the cryoscopic data. A chain length of about 100 maltose units for starch and about 40 for dextrin was concluded. The first separation of starch into amylose, which could be completely converted to maltose, and amylopectin, which was converted to dextrin, was reported in 1906 [2.16]. At that time starch and dextrin were considered to be colloidal substances, i.e. aggregates of small molecules, and the "chain lengths" reported were not accepted as representing true molecular weights.

2.3.3
Cellulose

Studies of the chemical structure of cellulose were even more difficult than those of starch due to its insolubility in all common solvents. The first isolation of cellulose by Payen [2.4] in 1839 has already been mentioned. The conversion of cellulose to glucose by acid hydrolysis was reported 20 years later. The relations between cellulose, starch and dextrin was a controversy for many years. The opinion among scientists varied from different states of aggregation of the same basic substance (a carbohydrate) to different isomeric states of the substances. Another problem was differences between cellulose of various origin.

Cotton fiber cellulose yielded only glucose when hydrolyzed while wood cellulose fibers contained several other sugar residues. Wood hemicelluloses as different substances in the plant cell walls were discovered and analyzed much later.

A basic chemical difference between starch and cellulose was resolved in 1901 by Skraup and König [2.17] who showed that cellulose was degraded to cellobiose which was different from the maltose derived from starch. The chemical structure of the cellulose molecule remained a controversy and several models of ring and chain structures were proposed. Strong evidence for chain structure of cellulose was given by Ost [2.18]. He used both reducing end group analysis and viscosity of cellulose in copperammonium solution to show that hydrolysis with acid caused a gradual decrease in chain length. Ost stated correctly that "hydrocellulose" was not a cellulose hydrate but cellulose degraded to low molecular weight, i.e. short chain length, by acid hydrolysis.

2.3.4
Lignin

"Incrusting material"‟ was removed from plant cell walls when Payen [2.4] first isolated cellulose by treatment with nitric acid in 1839. The term "lignin" was introduced in 1865 by F. Schulze (derived from "lignum", the latin word for wood). In spite of the large amounts of lignin in wood (20–30 %) and other plant materials and the large-scale processes for production of paper fibers from wood, there were few attempts to analyze the structure of lignin all through the 1800s. An exception was Peter Klason in Sweden. He devoted several decades of research to lignin, in particular the reactions of lignin in the sulfite process when lignin is removed as water-soluble lignosulfonic acids [2.19]. In 1897 Klason found that lignin is chemically related to coniferyl alcohol as a main component.

Klason continued his research, and in 1907 he published the idea that lignin in its native form is a large molecule which cannot be analyzed and characterized by organic chemical methods. Peter Klason was educated as an organic chemist. He concluded in 1917 that the coniferyl alcohol units were joined together by ether linkages. Klason's results were later verified by other scientists and they form the basis for research by which the structure of lignin in wood has been largely resolved [2.20].

Peter Klason's research is of general interest related to the emergence of the macromolecular concept although the results were overlooked at the time.

Klason showed clearly that coniferyl alcohol is a *monomer* which forms *large lignin molecules* held together with *ether bonds* which are normal chemical bonds. These are the elements of Staudinger's general concept of macromolecules presented in 1920.

2.3.5
Proteins

Compared with rubber, starch, cellulose and lignin, proteins have a much more complex structure. The history of protein research is well covered by Fenton [2.21] and only a short review will be given here.

A number of plant and animal proteins were studied during the early 1800s: albumin (from egg-white), casein (from milk), fibrin and serum albumin (from blood). Proteins were also found in plants but were for a long time considered different from animal proteins. It was gradually discovered that all proteins contain a certain number of different amino acids bonded together to polypeptides and released by hydrolysis. When Emil Fischer started his protein research around 1900, he was eventually able to synthesize a polypeptide chain containing 18 amino acid residues of glycine and leucin [2.22] with a molecular weight of about 1200. In the discussions of protein structure, ten times higher molecular weights were considered possible for the polypeptides, assumed to be aggregated to form the native proteins. The pH sensitivity and the denaturation and coagulation by heating in aqueous solution were taken as support for the aggregation theory of protein structure. The general opinion among scientists in the early 1900s was that proteins were colloids and not large molecules.

2.4
The Macromolecular Concept

2.4.1
Staudinger's Contributions

As described in the previous section, many scientists in the late 1800s and the early 1900s reported that native organic materials contained molecules of larger size than the regular organic substances which were synthesized, analyzed and described.

The molecules derived from native materials were considered to be of intermediate size. This was evidenced for Hevea rubber, starch, cellulose, lignin and proteins. Around 1900 colloid chemistry had emerged into a strong position among scientists.

Like inorganic colloids of metals, sulfur, clays etc., the native organic materials were generally considered to be aggregates of molecules held together by secondary forces and not by primary chemical bonds. The arguments for the aggregation theory were questioned and even considered controversial in discussions by some leading scientists. While most physical chemists defended the colloid concept, it was an organic chemist, Hermann Staudinger, who first clearly defined molecules of very long chains held together by normal chemical bonds. He became the leading champion for the macromolecular concept in the 1920s and 1930s.

Hermann Staudinger was born in Germany in 1881 and educated as an organic chemist. His first appointment was as "Dozent" (assistant professor) at the Technical University in Karlsruhe in 1907. His studies of ketenes and diazo compounds qualified him as a scientist, and in 1913 he was appointed as Professor of Organic Chemistry at the prestigious Federal Polytechnic Institute (ETH) in Zürich. Staudinger's interest in large molecules goes back to his years in Karlsruhe, 1907–1913. One of his students there, L. Lautenschläger, worked on polymerization and autooxidation of olefins. For the structure of the resulting rubber polymers, Harries' model of aggregated cyclic octadiene was accepted. Staudinger continued his research with other monomers and prepared polymers of formaldehyde, vinyl bromide and styrene. The results gradually convinced him that the monomers formed long chains held together by covalent bonds and were not associated ring structures as previously believed. Staudinger presented the idea of "hochmolekulare Verbindungen" (high molecular compounds) to the Swiss Chemical Society in 1917 in a lecture which was published two years later. A more extensive presentation of the macromolecular concept with experimental details was published by Staudinger in 1920 [2.23]. This year, therefore, is recorded in the literature as the initiation year of the macromolecular concept.

There were several problems with the macromolecular concept which Staudinger worked on during the 1920s. One problem was the end groups. In polyoxymethylene polymerized from formaldehyde, Staudinger argued correctly that both chain ends carry hydroxyl groups which could add new monomers.

For polystyrene the end group structure was a difficult problem. Straightforward polymerization of styrene would leave the end groups as open valence bonds, i.e., reactive free radicals. Staudinger offered two different explanations. One possibility was, according to Staudinger, that the two chain ends reacted by combination and formed large ring structures. Another possibility was that the chain ends would be "diluted" and lose reactivity due to the large chain length. The end group problem was eventually settled when the mechanism of polymerization was worked out: one chain end was formed by the initiation reaction and the other chain end by the termination. Both chain ends would then contain normal chemical bonds.

Staudinger's macromolecular concept was severely criticized and not accepted by many prominent scientists during the 1920s. With our present knowledge, the objections to Staudinger's ideas do not appear well founded. Some scientists flatly refused to believe that such a long molecular chain could exist. It would simply fall apart. That Hevea rubber and starch did not crystallize was interpreted as due to impurities and not to long chain structure. Other scientists argued that a molecule could not be larger than the crystallographic unit cell in the solid state, because they did not accept that a polymer chain could run through a sequence of adjacent unit cells. The concentration dependence of rubber and cellulose solutions was taken as support for increased association, i.e., larger micelles, at higher concentrations. Staudinger defended his macro-molecular concept in heated discussions which could even be hostile. Some colleagues advised Staudinger to abandon research on materials which could not be distilled or crystallized. With contempt, Staudinger's work was called "grease chemistry".

But Staudinger was a determined and courageous man. He had decided in 1920 to devote his research to "hochmolekulare Verbindungen"[']. After he had moved from Zürich to Freiburg in Breisgau in Germany in 1925, as Professor of Organic Chemistry, he continued his struggle to have his views accepted [3.24].

On some points he was forced to accept the views of his critical colleagues, including the explanation for the strong concentration dependence of macromolecules in solution. Staudinger interpreted this effect as due to increasing association of macro-molecules to irregular micelles, i.e., colloids. But Staudinger remained convinced that the aggregates contained macromolecules and not small molecules. Later research on macromolecules in solution by Flory and others has shown that concentration dependence is a normal phenomenon due to interaction between flexible chain molecules in solution. It was later found that stiff chain molecules can associate to bundles of parallel chains or chain segments ("Liguid Crystal Polymers") but this phenomenon gives a decreased viscosity of the solution at an intermediate concentration.

The macromolecular chain concept according to Staudinger's definition was generally accepted by leading scientists in the early 1930s. This structural principle meant [2.25] *that* macromolecules were not associated small molecules (colloids) but chain molecules held together by normal chemical bonds, *that* the chains have characteristic end groups with a structure related to the reaction by which the chain is formed, *that* the end groups can be analyzed and used for determination of the average chain length, *that* a macromolecular chain can run through many unit cells in a crystal lattice and also *that* chain ends can be incorporated randomly in the crystal lattice.

2.4.2
Supporting Evidence for the Macromolecular Concept

Staudinger was not alone in his struggle to have the macromolecular concept accepted. A colloid chemist at Uppsala University, The(odor)* Svedberg, the first professor of physical chemistry in Sweden, constructed the ultracentrifuge in 1923 to measure the size of very small colloidal particles by sedimentation in strong centrifugal fields. The sedimentation was observed optically during the spinning of the centrifuge.

In 1924 centrifugal fields of 5000 g were reached (g is the gravity of the earth at sea level). An attempt was made with this instrument to study the particle size of haemoglobin, the red protein in blood cells. Svedberg had expected to find a dispersion of aggregated polypeptides and amino acids according to Emil Fischer's view [2.22]. Much to his surprise, Svedberg found that the haemoglobin sedimented to an equilibrium which indicated a uniform molecular weight calculated to be 68 000 [2.26].

Svedberg concluded that haemoglobin consists of macromolecules of a uniform size and well defined structure. This was also found to be true for other soluble

*) He shortened his name from the given Theodor to "The", pronounced Te like in ten.

proteins and for vegetable polysaccharides from bulbs, studied by Svedberg with the ultracentrifuge. These results initiated macromolecular biochemistry as a research field and gave strong support for the macromolecular concept [2.27].

In 1928 a young organic chemist, Wallace H. Carothers, was hired by the Du Pont Company in Wilmington, Delaware, to be head of a research group to work on fundamental organic chemistry. Carothers had obtained his doctor's degree as a student of Roger Adams at the University of Illinois, the leading organic chemist in USA at that time. Already, before he came to Du Pont, Carothers had become interested in synthesis of long molecular chains by condensation reactions. His first work there was on polycondensation of dibromoalkanes like 1,10-dibromodecane with sodium at elevated temperatures. Carothers continued with preparation of polyesters from diols and dicarboxylic acids. Much higher molecular weights than those reported by Staudinger were obtained. It was discovered that these polyesters, both as melt and in solid state, could be drawn into strong flexible fibers. The aliphatic polyesters he prepared had too low a melting point (below 100 °C) to be useful as a synthetic fiber. After a few years at Du Pont, Carothers decided to synthesize polyamides as fiber material. After unsuccessful attempts with various amino carboxylic acids, Carothers prepared in 1935 a high-melting polyamide from hexamethylene diamine and adipic acid, which was easily drawn into strong fibers of high melting point (255 °C).

Three years later the new polymer was introduced on the market as ladies' stockings of the miraculous Nylon fiber, the "New Silk" which was widely hailed as the introduction to the new age of synthetic materials. Carothers' research produced another important polymeric material, the elastomer poly(chloroprene) which was marketed as Neoprene rubber with excellent resistance to oil and to ageing. At the same time Carothers was active in basic research and a strong advocate of the macromolecular concept [2.28].

In 1934, Paul J. Flory joined Carothers' research group and started his work on the theory of polymerization reactions, both by condensation (step-growth) and addition (chain-growth) polymerization.

The work of Carothers and Flory produced the definitive verification of the macromolecular concept. Carothers' polymer research is presented in a volume edited by H. Mark and G.S. Whitby [2.29].

2.4.3
The Established Macromolecular Concept

The macromolecular concept is now accepted and well established as the basis for a most active branch of chemical research. Macromolecules are usually polymers, i.e., they are formed from small molecules, monomers, which react to macromolecules with one or a few repeating units in their structure. Other macromolecules are formed by growth of small molecules to more irregular structures of high molecular weight (mass). The macromolecules may be linear chains, branched chains or networks in two or three dimensions. There is no upper limit for the molecular mass of a macromolecule. When multifunctional monomers react they form branched molecules which eventually become networks of infinite molecular mass.

Macromolecules are now the structural elements of polymeric materials which are rapidly increasing in produced amounts and specialized properties as construction materials. In biophysics and biochemistry with medical applications, macromolecules are the key elements and the basis for further progress. Biopolymers are carriers of our main biological functions including our mental processes. Biomedical polymers are essential as spare parts for our body, i.e., synthetic implants for joints, blood vessels, heart valves, eye lenses, hearing bones, etc.

Further progress in materials science and medical technology requires macromolecular research and development which is described in the following chapters of this book.

2.5
References

2.1   Morawetz H (1985). Polymers. The origins and growth of a science. John Wiley, New York.
2.2   Flory P J (1953). Principles of polymer chemistry. Cornell University Press, Ithaca, New York.
2.3   Törnqvist E G M (1968). In: Kennedy, J P, Törnqvist, E G M (eds). The polymer chemistry of synthetic elastomers. Wiley-Interscience, New York.
2.4   Payen A (1839). Compt Rend 8:51.
2.5   Sjöström E (1981). Wood chemistry. Fundamentals and applications. Academic New York, p. 104.
2.6   Kaufman M (1963). The first century of plastics, celluloid and its sequel, The plastics and rubber institute, London.
2.7   Baekeland L H (1909). Ind. Eng. Chem., 1, 149.
2.8   Faraday M (1826). Quart. J. Sci. Arts. Roy. Inst. Great Britain 21, 19
2.9   Harries C (1902). Ber. Deutsch Chem. Ges. 35, 3256.
2.10   Pickles S S (1910). J. Chem. Soc., 97, 1085.
2.11   Gough J (1805). Mem. Lit. Phil. Soc., Manchester [2], 1, 288
2.12   Guth E and Mark H (1934). Monatsheft der Chemie, 65, 93
2.13   Kirchhoff Cf C (1815). J. Chem., 14, 385 and 389.
2.14   O'Sullivan C (1876). J. Chem. Soc., 30, 125
2.15   Brown Cf H T and Millar J H (1899). J. Chem. Soc., 75, 286.
2.16   Maquenne L and Roux E (1906). Ann. Chem. Phys., [8], 9, 179.
2.17   Skraup Z H and König J (1901). Ber. Deutsch Chem. Ges., 34, 1115.
2.18   Ost H (1913). Ann. der Chemie, 398, 313.
2.19   Hägglund Cf E (1951). Wood Chemistry. Academic Press, New York, p. 181.
2.20   Sjöström Cf E (1981). Wood Chemistry. Fundamentals and applications, Academic Press, New York, p. 68 ff.
2.21   Fenton J S (1972). Molecules and Life, Wiley, New York.
2.22   Fischer E (1907). Ber. Deutsch Chem. Ges., 40, 1754.
2.23   Staudinger H (1920). Ber. Deutsch Chem. Ges. 53, 1073.
2.24   Staudinger H (1970). From Organic Chemistry to Macromolecules, Wiley-Interscience, New York. This book is a translation of H. Staudinger, Arbeitserinnerungen, A. Hüthig Verlag (1961), Heidelberg.
2.25   Morawetz H ref. 1, p. 92.
2.26   Svedberg T and Pedersen K O (1940). The Untracentrifuge, Clarendon Press, Oxford.
2.27   Rånby B (ed.), Physical chemistry of colloids and macromolecules. The Svedberg Symposium. (1987). Blackwell Scientific Publ., Oxford.
2.28   Carothers W H (1931). Chem. Revs., 8, 353.
2.29   Mark H and Whitby G S (Eds.), (1940). Collected papers of Wallace Carothers on polymerization, Interscience, New York.

Chapter 3

Polymers – A Growing Science

Bengt Rånby

3.1
Introduction

The previous chapter describes the emergence of the macromolecular concept, its formal definition by Staudinger in 1920 and its general acceptance around 1930. During the following 60 years, polymer science has grown to a large, diversified and advanced field of research. It is one of the most active and rapidly expanding branches in chemistry, physics, materials science, molecular biology and medicine. This chapter reviews the main advances in the wide area of polymer science.

3.2
Polymerization

The first more systematic study of polymerization was made by Hermann Staudinger in the 1920s and the early 1930s. For the mechanism of free radical polymerization he defined three reaction steps [3.1, 3.2]:

initiation – radical formation and addition of a radical to a monomer molecule,
propagation – chain growth by subsequent monomer addition to chain ends, and
termination – chain end reactions which stop the chain growth.

In 1937, Paul J. Flory, working with W.H Carothers at Du Pont, postulated the *chain transfer mechanism* in radical polymerization [3.3]. The growing radical chain end abstracts, for example, a hydrogen, which terminates the chain. The new radical initiates a new chain. Based on this concept, the average chain length of the polymer can be controlled by adding to the polymerizing monomer a chain transfer agent from which hydrogen is easily abstracted. Chain transfer reactions have also been found for ionic and coordination polymerizations.

W. H.Carothers made several significant contributions to polymer science during his ten-year period at Du Pont (1928–37). He prepared linear polyesters and polyamides of high molecular weight. He discovered the formation of strong fibres by cold-drawing of a melt of a linear polymer. During this work Carothers concluded that the step-growth polymerization was a regular condensation reaction in which the reactivity of the end groups of growing chains could be assumed to be independent of the chain length [3.4] Carothers' synthesis of 6,6-nylon led to the production of a remarkable textile fiber, a strong film and a tough plastic material. He also synthesized

Okamura, Rånby, Ito (Eds.): Macromolecular Concept and Strategy
© Springer-Verlag Berlin · Heidelberg 1996

poly(chloroprene) which gave an oil and weather resistant rubber. These achievements greatly inspired the industrial development of synthetic polymers all over the world. For the general public, 6,6-nylon fibres from poly(hexamethylene-adipamide) initiated the era of synthetic polymers when nylon stockings were first introduced and marketed in the USA in 1940.

During his years at Carothers' laboratory in 1934–38, Paul J. Flory derived the *chain length* (or molecular weight) and the *chain length distribution* for polymers by regular step-growth polymerization. Flory continued his theoretical studies with chain-growth polymerization and derived Gaussian distributions for polymers formed by free-radical polymerization and Poisson distributions for polymers formed without chain transfer and chain termination [3.5].

Ionic chain polymerization started as an industrial development of butyl rubber in Germany (1932). The polymer was produced by polymerization of isobutene at low temperature initiated by the addition of borontrifluoride. The polymerization mechanism was established in the 1940s as a *cationic reaction* with the growing chain end positive. The BF_3 requires the presence of small amounts of water or alcohol as a cocatalyst to form a complex. Dissociation of the complex gives a proton which adds to the monomer and starts cationic polymerization [3.6].

Anionic chain polymerization was also established in the 1940s and applied to monomers like methacrylonitrile, butadiene and styrene with sodium, or triphenylcarbions as initiator. An important advance was made by Karl Ziegler in 1950 [3.7]. He polymerized butadiene with butyl lithium as catalyst and obtained a polymer with high degree of 1,4-addition and prevailing cis-structure of the vinylene groups in the polybutadiene chains.

The research of Ziegler's group in the 1950s lead to the discovery of *coordination polymerization of olefins* with the complex metal-organic catalyst of aluminium alkyl/titanium chloride [3.8]. This research started with the observation that aluminium trialkyl can add ethylene at room temperature and low pressure and form short alkyl chains (1953). With titanium chloride present, long chains of linear polyethylene were formed. Using this catalyst, Giulio Natta discovered in 1955 that propylene polymerized to a crystalline polypropylene containing chain segments of stereoregular structure [3.9]. The important discovery of the Ziegler-Natta catalysts for *stereospecific polymerization* of olefins actively promoted extensive research in the new field of stereoregular polymers of various other monomers and studies of the stereochemistry of polymer chains.

Ring-opening polymerization was first applied by Schlack in 1941 for the synthesis of 6-nylon from caprolactam [3.10]. This seven-membered ring is opened by acid or base catalysts to form the linear polycaprolactam or 6-nylon. Ring-opening reactions had previously been studied for the synthesis of polypeptides from amino acid – N – carboxyanhydrides [3.11] and were later developed for polyesters and polyanhydrides from ring-closed lactones and cyclic ethers [3.12]. One advantage of ring-opening polymerization reactions is that they give polymers with only small volume changes because the number of chemical bonds is not changed in the process. In addition, the ring-closed monomers are usually easy to prepare and purify.

It was discovered by Michael Szwarc in 1956 that certain anionic polymerizations had no spontaneous termination and no chain transfer reactions [3.13]. All chains started to grow at the same time when the initiator was added and the chains continued to grow until all the monomer was consumed. The chains start to grow again if a new monomer is added. The chains of these *"living polymers"* all grow simultaneously for the same length of time and have, therefore, a very narrow chain length distribution, i. e., a Poisson distribution as already calculated by Flory in 1940 [3.5]. This mechanism has made it possible to synthesize polymers with narow chain length distribution and copolymers containing segments (or blocks) of different structure by adding one monomer (A) after another (B), e. g. to AAA . . . ABBB . . . B or AA . . . ABB . . . BAA . . . A block copolymers, which are commercially produced as thermoplastic elastomers. If the end segments (AAA. . .A) have high softening temperature and are incompatible with the middle segments (BB. . .B), the A-segments form physical crosslinks between the more flexible B-segments with low softening temperature.

3.3
The Molecular Weight (Mass) of Macromolecules

An early and very important determination of the molecular weight (or mass) of a macromolecule was made by Svedberg in 1925. He studied sedimentation equilibrium of haemoglobin in the ultracentrifuge, and discovered that these protein molecules have a uniform molecular weight [3.14]. This unexpected result was later verified for haemoglobin and other proteins by measuring the rate of sedimentation in an ultra-centrifuge with higher centrifugal fields combined with measurements of diffusion rates. These measurements of molecular weight were accepted as experimental evidence for the existence of macromolecules. The results showed that proteins in solution are macromolecules and not gels of aggregated amino acids as previoulsy believed. Svedberg's work initiated biochemical and biophysical research on a macromolecular basis.

Viscosity measurements for substances in solution were applied in 1910–20 for natural rubber, dextrin (hydrolyzed starch), gelatin and cellulose derivatives. The high viscosity of the solutions which further increased with increasing concentration was interpreted at this time as due to increased association of small molecules. Another interpretation was given by Staudinger in the 1920s as described in the previous chapter. He argued that solution viscosity measured chain length (or molecular weight) as a linear function of solution viscosity [3.15]. Staudinger maintained the view into the 1950s that macromolecules in solution were rigid rods. Other scientists like Herman Mark and Werner Kuhn developed theories for solution viscosity based on the assumption of flexibility of chains as due to more or less free rotation around the chemical bonds [3.16]. These theories were supported by experimental evidence [3.17]. The solution viscosity increased as an exponential function M^a of the molecular weight M with $0.5 < a < 1$ according to the Mark-Houwink-Sakurada equation.

$$\lim(\eta/c) = KM^a \ (c \rightarrow o)$$

where $\eta_{sp} = \eta_{rel} - 1$ at a polymer concentration c and K is a parameter related to the polymer / solvent interaction. The M-values from viscosity measurements are convenient to obtain but they are relative and have to be calibrated with other methods, e. g., end group analysis or osmotic and ultracentrifugal measurements for determination of a- and K- values.

In relation to the interpretation of viscosity measurements, extensive theroretical studies were made on the properties of polymer chains in solution. It was found by Mark in 1930 that rigid rodlike chains would give an increase in solution viscosity proportional to the square of the chain length [3.18]. For flexible chains forming "random coils", Kuhn developed a polymer chain model which led to a relation of $\eta \, sp/c$ proportional to $M^{2/3}$ which was in reasonable agreement with experiments for several polymers [3.19].

Light scattering measurements became an important method for molecular weight determinations in the 1940s and 1950s. It was pointed out by Ostwald in 1930 that the turbidity of a protein solution was due to the scattering of light (the Tyndall effect) from the protein molecules with refractive index different from the solvent [3.20]. The theoretical relation between the Tyndall effect and the osmotic pressure for colloidal particles in solution was first treated by Raman [3.21]. The light scattering method for molecular weight measurements was introduced as a useful tool in polymer research by Debye in 1944 [3.22] and further developed by Zimm [3.23]. An advantage of the method is that the light scattering increases with increasing molecular weight. This makes light scattering useful for measurements of high molecular weights and an excellent complement to osmotic pressure which is proportional to $1/M$ and is therefore most useful for polymers of low and medium molecular weights (M).

Gel permeation chromatography (GPC) is another important method for molecular weight determination. It was originally initiated as gel filtration by Porath and Flodin [3.24] and based on the size of the polymer molecules in solution. The polymer solution flows through a porous bed of crosslinked gel particles with pore size of the same magnitude as the polymer molecules to be analyzed. Separation occurs because large molecules flow rapidly without entering the network of the gel particles while small molecules enter the network pores and flow slowly. By analysis of the effluent, a distribution of the molecular weights is obtained for the polymer. The measurements are calibrated with standard samples of polymers with known molecular weight.

3.4
The Physical Chemistry of Polymers

In the period 1942 to 1970, polymer research was established as a mature field of science. Due to the extensive contributions of Paul J. Flory the period has been called the *"Flory era"*. Flory's initial work on polymerization in 1934–40 has already been treated. The "non ideality" of a polymer solution was observed and studied for osmotic measurements in the 1930s but no interpretation was known. In 1942 Maurice Huggins and Paul Flory independently published a *thermodynamic theory of polymer solutions* which gives a quantitative interpretation of the non-ideality of a polymer solution [3.25, 3.26]. They derived an equation for the solvent activity a, as a function of the volume fractions of solvent (v_1) and polymer (v_2), the number (x) of lattice sites in the solution occupied by the polymer chain and an interaction parameter (χ) of polymer and solvent where RT $\chi\, v_2^2$ is the molar enthalpy of dilution:

$$\ln a_1 = \ln v_1 + (1 - 1/x)\, v_2 + \chi\, v_2^2.$$

This equation, which is called *"the volume fraction formula"* of polymer solutions, forms the basis for further developments of polymer physics, e.g., the theory for osmotic pressure and light scattering and their concentration dependence in polymer solutions and the properties of polymer gels and polymer melts [3.27]. The volume fraction formula is one of the fundamental laws of polymer science.

The *elasticity of rubber* as related to the end-to-end displacement and the retractive force of a flexible polymer chain was initiated in a statistical theory by Eugene Guth and Herman Mark in 1934 [3.28]. These calculations were later developed into a theory of rubber elasticity by Flory and Rehner [3.29]. They derived a function of the retractive force for a deformed rubber as proportional to the number of "subchains" between the crosslinks per unit volume, the temperature, and the function ($\alpha - I/\alpha^2$) where α is the relative extension of the rubber sample.

The *shape of the polymer molecules* for flexible polymer chains in solution was first calculated by Werner Kuhn in the 1930s. The theoretical work was continued by Flory in 1948, who developed a theory for *the excluded volume effect*, i. e., the effect on the dilute solution properties that a polymer coil tends to exclude other polymer molecules from entering the volume of the coil. This effect includes segments of the same polymer coil which extends the volume of the coil. This work on polymers in solution led to extensive theoretical treatment of solution properties of polymer chains in different solvents. Flory defined in 1949 a *θ-solvent* where the interactions between the chain segments and the solvent molecules give the chains unperturbed dimensions. In a θ-solvent the polymer chains are in a conformation equilibrium and the end-to-end distance of the coiled chain is proportional to the chain length [3.30]. Flory based a theory of polymer chain conformation in an amorphous bulk polymer on his theory of θ-solutions and predicted that the shape of the polymer coils in the bulk phase should be the same as in a θ-solvent. Using neutron scattering of a deuterated polymer dissolved in the original polymer, Flory's prediction was verified in the 1970s.

3.5
The Properties of Polymeric Materials

Polymeric materials have *viscoelastic properties*, i. e., their modulus has two components, *elasticity* which is linear with deformation and *viscosity* which depends on the rate of deformation. The mechanical properties are well described by the theory of viscoelasticity which was first developed by Turner Alfrey [3.31] and brought to completion by John D. Ferry [3.32]. Polymeric materials show two transitions when the temperature is increased: a *glass transition* (T_g) due to segmental motion in amorphous polymer and a *melt transition* (T_m) due to melting of crystalline polymer. An amorphous polymer is usually brittle at temperatures below T_g. A crystalline polymer normally shows toughness at temperatures between T_g and T_m. Both transitions are measured by differential thermal analysis (differential scanning calorimetry, DSC) and interpreted thermodynamically.

It was discovered in the 1950s that polymer chains may crystallize by *chainfolding* and form thin lamellar crystals [3.33]. Extensive stretching of a crystalline or amorphous polymer near the melting point gives a fiber or sheet structure with the chains well oriented in the stretching direction. Very strong fibers are formed from crystallizing linear polymers by this process.

Rubber materials of high elasticity are polymers crosslinked to chain networks and used at temperatures well above the glass transition (T_g). Crosslinking is also applied to thermoplastic materials to increase the deformation temperature and the mechanical strength. Thermoset polymers are crosslinked during processing.

3.6
Analysis of Polymer Chain Structure

The methods for spectroscopic analysis of polymer chain structures have made spectacular progress during the last few decades. Using *infrared spectroscopy*, the absorption lines of the spectra of polymers in solution, in melt or in solid state, are recorded and assigned to specific resonance vibrations of polar groups in the main chain or in sidegroups [3.34, 3.35]. Using laser beams and computerized spectrometers, Fourier transform infrared spectra (FTIR) can be rapidly recorded, resolved and assigned to specific groups. For nonpolar groups in a polymer, e. g., C-C bonds, *Raman spectra* can be recorded and used as a complement to IR spectra.

The most powerful method for structure analysis of polymers is *nuclear magnetic resonance* (NMR) spectroscopy [3.36]. It is based on the resonance vibration of polar nuclei with spin, e. g, in ^{1}H, ^{13}C, ^{14}N, ^{19}F and ^{31}P. The modern instruments are so sensitive and selective that a total analysis of a chain structure like branched polyethylene is possible from one multiple recorded IR spectrum. This includes branchpoints, end groups, length of side chains and the molecular weight of the polymer molecule.

3.7
Liquid Crystal Polymers (LCP)

The liquid crystal phenomenon was observed for cholesterol esters with rodlike molecules more than 100 years ago [3.37]. These substances had two melting points between which the liquid showed iridescent colors and optical birefringence. These materials were called "flowing crystals", and "crystalline liquids". In the 1920s the concept was gradually extended to synthetic oligomers with stiff chains of increasing chain length. Successful synthesis of *liquid crystal polymers* (LCP) was reported in the 1950s by a research group at Courtaulds in England [3.38]. They prepared poly(γ-benzyl – L – glutamate) as a fiber polymer and observed the liquid crystal phases while spinning this polymer (PBLG) to fibers from solution. Theories for the formation of the LC lyotropic phase was developed by Flory in 1956 [3.39]. In addition, Flory made calculations of the viscosity of rodlike polymer chains in solution and predicted viscosity minima for an intermediate concentration of polymer and for an intermediate solution temperature [3.39].

The first large scale product of LC polymers was poly(p-phenylene-terephthalamide) (PPTA) which is spun into strong and high melting fibers under the trade name Kevlar. Many LC polymers are now synthesized containing stiff chain segments called *mesogenic groups* connected with flexible chain segments. The mesogenic segments can also be side groups connected with the main chain with flexible segment [3.40].

3.8
Electrically Conducting Polymers (ECP)

Organic polymers in commercial use have high electric resistance, i. e., they are insulating materials. The nonpolar polymers also have low electric loss factors. In 1977 it was discovered that polyacetylene, a polymer with conjugated double bonds along the chain, could by "doped" by electron donating or electron accepting reagents. The unpaired electrons on the chains are mobile and make the polymer electrically conducting with a resistance in the same region as metals such as iron and mercury. This discovery initiated extensive research in laboratories all over the world. A whole group of ECP materials with various doping are now known. A few applications for ECP have been found, e. g, as shields for electric fields, as material for rechargeable batteries and in thermo- and photoelectric devices. The ECP development was reviewed at a Nobel Symposium in 1991 and the papers presented there published as a symposium volume in 1993 [3.41].

3.9
Conclusions

This chapter has reviewed the growth of polymer science as a discipline.

The methods of polymerization by radical, ionic, coordination and ring-opening mechanisms have been described.

The methods of molecular weight (mass) and molecular weight (mass) distribution measurements have been reviewed in relation to the physical chemistry of polymer chains in solution, in gels and in the solid state.

The mechanical properties of polymeric materials have been related to chain structure, chain mobility, crystallization of polymer chains and the thermodynamic transitions of amorphous polymer phase (glass transition T_g) and crystalline phase (melt transition T_m).

Two more recent developments in polymer science are the liquid crystal polymers (LCP) with stiff chains or chain segments which show molecular orientation and crystalline order in the liquid state, and the electrical conductive polymers (ECP) which have conjugated chain structures and, after doping, give electrical conductivity of the same order as metals. These types of polymers have introduced new branches of polymer science and offer new applications as optical, magnetic and electrical materials in advanced technologies.

References

3.1 Staudinger H, Frost W (1935) Zeitschrift für Physikalische Chemie B 29: 235
3.2 Staudinger H, Steinhofer A (1935) Annalen der Chemie 517: 35
3.3 Flory P J (1937) J Am Chem Soc 59: 241
3.4 Carothers W H (1931) Chem Revs 8: 353
3.5 Flory P J (1940) J Am Chem Soc 62: 1561
3.6 Price C C (1944) Ann N. Y. Acad Sci 44: 351
3.7 Ziegler K, Eimers E, Heckelhammer H and Wilms H (1950) Ann der Chemie, 567: 43
3.8 Ziegler K, Holzkamp E, Breil H and Martin H (1955) Angew; Chemie 67: 541
3.9 Natta G, Pino P, Corradini P, Danusso F, Mantico E, Mazzanti G and Moraglio G (1955) J Am Chem Soc. 77, 1708
3.10 Schlack P (1941) U. S. Pat. 2,241, 321
3.11 Leuchs H (1906) Ber. 39, 857
3.12 Lundberg R D and Cox E F, in "Ring opening polymerization" (Frisch K C and Reegan S L, eds), 1969. Chapt. 6, Marcel Dekker, New York
3.13 Szwarc M (1956) Nature 174, 1168 and Szwarc M, Levy M and Milkovich R (1956) J Am Chem Soc. 78, 2656
3.14 Svedberg T and Fåhraeus R (1926) J Am Chem Soc. 48, 430
3.15 Staudinger H and Heuer W (1930) Ber. 63, 222
3.16 Kuhn W (1938) Kolloid-Z., 76, 258 and (1938) Angew. Chem. 51, 640
3.17 Fikentscher H and Mark H (1939) Kolloid-Z, 49, 138
3.18 Mark H (1930) Kolloid Z, 53: 32
3.19 Haller W (1931) Kolloid-Z. 56: 257
3.20 Ostwald W (1930) Kolloid-Z. 53, 42
3.21 Raman C V (1927) Nature 120, 158
3.22 Debye P (1944) J. Appl. Phys. 15, 338
3.23 Zimm B H (1948) J. Chem. Phys. 16, 1099
3.24 Porath J and Flodin P (1959) Nature 183, 1657 See further a review in Johnson J F and Porter R S (1968) (eds.), "Analytical Gel Permetion Chromertography", Wiley-Interscience, New York.
3.25 Huggins M L (1942) J. Phys. Chem. 46, 151
3.26 Flory P J (1942) J. Chem. Phys. 10, 51
3.27 See further Flory P J (1953) "The principles of polymer chemistry", Cornell University Press, ITHACA, New York
3.28 Guth E and Mark H (1934) Monatsheft Chemie 65, 93
3.29 Flory P J and Rehner J (1943) Jr., J. Chem. Phys. 11, 512
3.30 Flory P J (1949) J. Chem. Phys. 17, 303
3.31 Alfrey Jr T (1948) "Mechanical Behavior of High Polymers", Interscience, New York
3.32 Ferry J D (1961) J. Am. Chem. Soc. 72, 3746 (1950) and "Viscoelastic properties of polymers", J. Wiley, New York
3.33 A review in Sharples A (1966) "Introduction to Polymer Chrystallization", E. Arnold, London
3.34 Zbinden R (1964) "Infrared spectrocopy of high polymers", Academic Press, New York
3.35 Hummel D O (1966) "Infrared Spectra of Polymers", Wiley-Interscience, New York
3.36 Bovey F A (1969) "Polymer conformation and configuration", Academic Press, New York
3.37 Reinitzer F (1888) Monatsheft Chemie, 9, 421
3.38 Robinson C (1956) Trans. Faraday Soc. 52, 571
3.39 Flory P J (1956) Proc. Royal Society, London, A 234, 60, 73
3.40 For a review, cf. Donald A M and Windle A H (1992) "Liquid crystal polymers", Cambridge Univ. Press
3.41 Salaneck W R, Lundström I and Rånby B (1993) (eds) "Conjugated polymers and related materials", Oxford Univ. Press, Oxford, UK

Chapter 4

A New Trend in Polymer Science and Technology for Humanity

Seizo Okamura

4.1
Introduction

Things might be evaluated both by technique (truth) and by arts (beauty), just as the technical merit and the artistic impression in figure skating. Even in mathematics, it is often said that beauty means truth, as in evaluating mathematical formulae. Thus, in general, any concept must be true *and* beautiful, and this will be equally applied to any strategy of action.

In this chapter, I would like to consider the way to approach arts from technique, specifically in terms of polymeric materials. My discussion will be confined to this one-way process, from technique to arts alone; another process in the opposite direction, from arts to technique, is examined in detail by Yoda in Chap. 6 of this volume.

In discussing the approach of polymeric materials from technique to arts, I start with the "property" as a basis for any science and technology (Fig. 4.1, the right half). Following is the "functionality" of polymeric materials, which is discussed in some detail, and then comes the "sensibility" in an active meaning or the "sensitivity" as a passive counterpart. Herein the stream from property through functionality and sensitivity to arts is reviewed on the basis of my research experience [4.1]. A decade ago, for example, we studied biocompatibility in detail, which we regarded as one of the functionalties of polymers that are the pre-stages to reaching the sensibility towards living body and finally towards humanity itself.

In the last part of this chapter, I will examine rather briefly the relationship between technique and arts in two ways as illustrated in Fig. 4.1: namely, one way is

Fig. 4.1 Approaches from technique to arts and vice versa

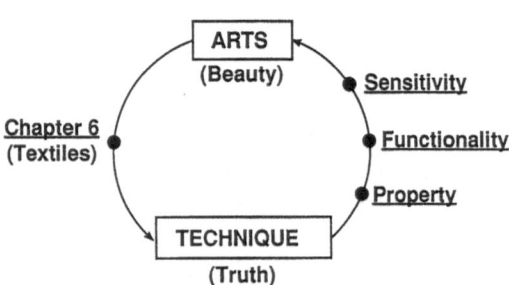

artistically "downwards", from arts to technique (the left half), and the other is technically "upwards", from technique to arts (the right half).

4.2
Property of Polymeric Materials as the Primary Performance

Our early research during the period 1937–40, or about 55 years ago, began with the measurement of the physical properties of textile fibers, such as the breaking strength and elongation of artificial wool-like fibers made from soybean proteins [4.2]. That was a part of our continuing efforts to obtain exact data of the physical properties of polymers, particularly man-made fivers. However, we soon found the experimental values to depend, albeit slightly, on the conditions of our measurements. It is also known that physical property data vary with the analytical methods (mechanical, chemical, etc.) that are used for particular experiments. Thus, property, or one of the answers (A) for experiments, depends on the method or the condition of a measurement, or one of the questions (Q) to find a particular answer.

Such an interdependence between Q and A is rather surprising, though frequently noticed, because, for example, polymeric materials are regarded as a part of "Nature", which is the general and ultimate subject of natural sciences; almost subconsciously, we natural scientists assume that "Nature" gives the same A for a particular subject (physical property, etc.) regardless of the type of Q. Therefore, the interdependence of Q and A as we observed is an essential problem in science and technology, and it may be related to more philosophical problems like those between "existence" and "consciousness" or between human "body" and "mind" (cf. Chap. 11).

4.3
Functionality as the Secondary Performance

In the practical usages of polymeric materials, not a single but a set of their physical properties are often considered collectively, and the central focus in research and development is to optimize all of these properties. In such a process, a single Q can give several A, in one case, but in another, several Q can lead to a single A. Equally important, a particular Q sometimes results in different A depending on the changes in time and phase. We often consider the time-dependent properties such as delayed response to be "active" as an A for a Q, and the time-independent properties to be "passive". More recently, these complicated time-dependent properties seem to be termed fashionably as "functionality" (active term), which corresponds to the secondary performance of materials in my context (Fig. 4.1), in an attempt to distinguish it from the primary performance called "property" (passive term).

Generally speaking, the term "property" is used primarily for non-living subjects and is governed by the spatial axis (frame) alone, whereas the term "functionality" is originally for living or biological substances, which is now controlled by not only the spatial but also the time axes. Recently, our conceptual vector from "property" to "functionality" (Fig. 4.1, right half) has apparently been exaggerated or further extended, and the "functionality" tends to be mixed in usage with the term "sensitivity" or "sensibility", which is the third stage (tertiary performance) in the approach from technique to arts (Fig. 4.1). Compared with the "functionality", the term "sensitivity" is originally applied to living substances or more strictly to human beings alone. In contrast to the "property" and the "functionality", the "sensitivity" involves as many as three axes, spatial, time, and memory axes. Herein "memory", in turn, involves a "negative time scale" in the action of the human brain.

To summarize, despite some confusion in definition in the current terminology, the primary, secondary, and tertiary performances of polymeric materials and other substances are called "property", "functionality", and "sensitivity", respectively. They correspond, in this order, to space (S), space + time (S + T), and space + time + memory (S + T + M) [4.1]. Strictly, the S axis (frame) is for non-living substances; the S + T axes for living bodies (particularly simple ones like plants); and the S + T + M axes for human being (and perhaps advanced spezies that have as yet not developed). The "functionality" is also considered for such materials as "shin gosen" (new synthetic fibers), as discussed in Chap. 6.

4.4
Sensitivity as the Tertiary and Final Stage

Our research interest during the years 1970–75 was focused on the biomedical applications of polymers. For example, polymeric membranes coated with solubilized collagen for artificial tracheas were examined in terms of the intimate adhesion *in vivo* to the native collagen of body tissues [4.3]. The investigations were mainly directed to find suitable *in-vivo* conditions where the biological degradation of the solubilized collagen layer of the artificial organ is well balanced with the regeneration of native collagen in the surrounding body tissues. We found the rate of the biodegradation to be readily controlled by the extent of the radiation-induced crosslinking of the solubilized collagen macromolecules [4.4].

Thus, from the early period (1937–40; Sect. 4.3) to the first half of the 1970s, our research efforts shifted from the usual physical properties of textiles (a non-living aspect) to the special functionality of collagen polymers (a living aspect). In parallel with these research activities that were originally directed to the implanted artificial organs and their surface layer of solubilized collagen (or a non-living side), I developed some mental analogy that related the "property" of usual synthetic polymers to the "functionality" of biological macromolecules. Simultaneously, the idea of the "sensitivity" *came* into my mind, but in this case it was of course focused on native collagen, biological molecules regenerated in body tissues (or a living side). In this instance, therefore, the "functionality" of the solubilized collagen layer, as the "front side" of the tissue surface, is intimately "fused" to the "sensitivity" in the "back side" of the organ surface [4.5]. The fusion of two kinds of collagen molecules, one (the coating layer) from the artificial organ and the other from the body tissue, is thus an important factor that controls the adhesion between them (see also Chaps. 10 and 11).

In those days, some personal interest was held by me on the relationships between the "part" and the "whole" (entire body) of artificial organs systems, or more in

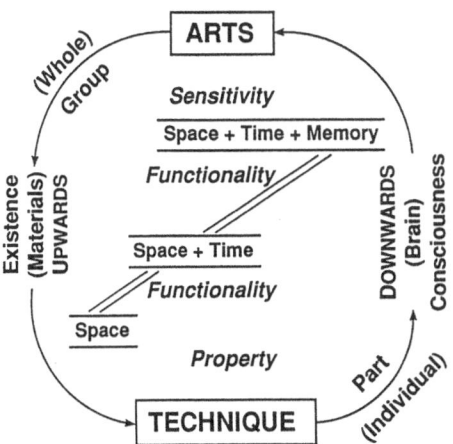

Fig. 4.2 Relationship between technique and arts through brain and materials

general, the relationships between an "individual" (*Teil* in German) and a "group" (*Ganz*) to which the former belongs in our way of thinking [4.6]. Consider an integer such as "1", for example. To this number, we may think that two slightly different numbers such as 1.0001 in its "front side" and 0.0009 in its "back side" are indeed fused [4.7], just as in the fusion of "functionality" and "sensitivity" mentioned above.

To visualize the conclusion of this section, Fig. 4.2 presents a diagram, which I would call a "pilled-up image of the entire relationship between technique and arts".

4.5
Functionality (Front Side) and Sensitivity (Back Side) in the Fusion of Technique and Arts

Today we have some troubles in the relation between advanced technology and public society, the problem called "public acceptance", such as those concerned with nuclear power generation and highly expensive medical treatment [4.8]. Regrettably we have not yet found solutions for these serious problems and, pessimistically, perhaps solutions might never be found which enable us to reconcile the "functionality" (F) or capability of advanced technology with the "sensitivity" (S) of the general public and society. In contrast, in the usual range of arts, similar "fusion" or "harmony" between F and S may be achieved more readily. For example, musicians can compose symphonies (active as F) in such a way as to make them, when performed, acceptable and comfortable (passive as S) in harmony with the "sensitivity" of the composers themselves and the general public. Namely, a kind of feedback and fine adjustment between F and S is highly possible [4.9]. However, as discussed in Chap. 10, the "fusion" of F with S at the management level in the chemical and materials industry (as F) in relation with the general public (as S) may equally be possible.

Just as important, the macromolecular concept could play an important role in connecting the "property" (non-living), the "functionality" (living), and the "sensitivity" (human) together and also in fusing "technique" (as F) with "arts" (as S). Listed below are several examples reported in the literature that show the achievements of such "fusions".

a) Professor Jacquelline Belloni-Coflar and her co-workers [4.10] have recently reviewed the reactions and physicochemical processes in photographic development. They clearly showed the importance of silver nuclei, formed by pulsed-light irradiation, which in turn act as a potential catalyst for the subsequent dark reactions to generate silver-black colloidal particles. This finding is a typical example of the "transformation" of a "property" to a "functionality", in the sense that the photographic emulsion plate (with a good "property") possesses a capability ("functionality") to respond actively (to generate a catalyst) to the irradiated light pulse.

Thus, these authors pointed out that "ever since the development reactions in photographic process, achieved by Louis Daquerre (1839), the key step of photographic reactions has benefitted from innumerable empirical improvements", and then they emphasized that "[the] recent physicochemical basis has been fully understood" and thus that "this enormous gain of research results in the photographic sensibility has led to the technology of emulsion having much finer particles and of much shorter time-exposure by several orders of magnitude." In my view, this is a typical example of the "fusion" between "technique" (the functionality of emulsion plates) and "arts" (the sensibility of finer and shorter-exposed photographs).

In the same context, one should also refer to the work of R. D. Michell, W. T. Nebe, and W. H. Hardam [4.11], who obtained for the first time a transparent polymeric matrix film in which precursor molecules of a photosentisizer are dispersed. The precursor was designed to be sensitive to light in a specific range of wavelengths. The first shot of irradiation at a particular wavelength thus generates a one-direction pattern of the photosentisizer derived from the dispersed precursor. The second irradiation at another wavelength then triggers the photosensitization to form a pattern in another direction. Only the doubly irradiated spots in the matrix can give final stripped lithographic patterns.

Similar studies in lithography have also been reported by H. Ito and C. G. Willson [4.12]. The IBM researchers made photoresist systems where a catalytic chain reaction can be initiated by an acid that is generated from a precursor (an onium salt) upon irradiation of light.

These two cases, discussed above, give examples of so-called "two-step amplification" reactions. In enzyme-catalyzed reactions, similarly, some protected chemical groups, originally inactive, are deprotected by a stimulation prozess to generate a catalytically active site. Thus, some enzymes can be activated only by such a two-step process. The incorporation of two kinds of molecules into polymer matrices (as in the first two cases) and the removal of protective groups by some stimulus (as in the third case) could be answers for constructing polymeric materials that could respond to various stimuli, or in other words, polymeric materials where the "functionality" and the "sensitivity" might suitably fuse.

b) The Society of Polymer Science, Japan once issued a special issue (November 1990) of its journal, Kobunshi (High Polymers), entitled "Kansei to Kobunshi" (or

"Sensitivity and Polymers") [4.13]. In the first article of this issue, M. Okamoto, Toray Company, defined the term "kansei" (sensitivity) along with some relevant general remarks. The other contributions gave specific consideration to "kansei" in various disciplines, including textiles (S. Yamaguchi, Kuraray Company), musical instruments (Y. Murase, Yamaha Company) cosmetics (M. Tanaka and S. Kumagai, Shiseido Company), hollography (Saito, Dainippon Printing Company), and taste perception (Y. Komata, Odawara Women's College). Since the appearance of this issue, apparently, Japanese polymer scientists and engineers have been strongly influenced by "kansei", a new concept (macromolecular concept) relevant to polymers.

Another new trend in the Japanese polymer industry is the recent commercialization of "shin gosen" (new synthetic fibers), which are finely fabricated polyester synthetic fibers that parallel natural silk in many aspects (see Chaps. 6 and 7). At least in Japan, these two trends indicate that polymer scientists and engineers increasingly pay attention to the "artistic" aspects of polymeric materials, which are of course controllable by the fabrication technique, i. e., this example also presents another way to "fuse" technique with public humanity.

c) A famous classical article should be mentioned here, which was published by G. T. Fechner in 1873, [4.14] entitled "Experimentelle Aesthetik." For the evaluation of paintings, he proposed an empirical formula that takes into consideration such factors as form (balance), color (harmony), and fineness (details), and then compared his results with the ratings [from 0 (worthless) to 5 (most excellent)] by professional art critics. As expected, his formula (as F) was too simple to give reasonable evaluations that agree with critics' ratings (as S). In contrast, for textile fibers we now have an objective and quantitative experimental method that parallels human perception or sensitivity, as described by Kawabata in Chap. 7.

4.6
References

4.1   This chapter is based in part on my essay on the "macromolecular concept" published in a series, "Rakuhoku Kei Ei (Reports from Northern Kyoto)" (hereafter K. Ei), in a monthly Japanese journal Kobunshi Kako (Polymers and Polymer Processing, hereafter K. K.): "Property, Functionality, and Sensitivity" (K Ei No. 120). Okamura S (1991) K K 40 (5): 223–224. A part of this chapter was also presented in the autor's lecture, "Polymer Science and Arts", at the 2nd Pacific Polymer Conference. October 1991, Otsu, Japan

4.2   Sakurada I, Okamura S (1938). X-genographic study and physical properties of artifical silk. Jinken Kai (World of Artifical Silk) 6 (1): 1

4.3a Yamada H, Yamada K, Kumagai H, Hino T, Okamura S (1978). Immobilization of ß-tyrosinase cell with collagen. Enzym Eng 3: 57–61;

4.3b Yamada H, Shimizu S, Tani Y, Hinto T (1980). Synthesis of coenzymes by immobilization cell systems. Enzym Eng 5: 405–411;

4.3c Shimizu Y, Abe R, Teramatsu T, Okamura S (1977). Study on copolymers of collagen and a synthetic polymer. I. Experimental study on bio-compatibility of laminar copolymers of collagen and a synthetic polymer. Biomat (Med. Div., Art. Org.) 5 (1): 49–66;

4.3d Miyamoto Y, Matsunobe S, Kato H, Shimizu Y, Abe R, Teramatsu T, Okamura S, Hino T (1978). Tracheal reconstruction using collagen-plastics copolymers, Jinko Zoki (Art. Org.) 7 (1): 165–167;

4.3e Yamada H, Asano Y, Hino T, Tani Y (1979). Microbial degradation of acrylonitrile, J Ferment Tech 57 (1): 8–14;

4.3f Miyamoto Y, Shimizu Y, Matsumoto S, Kato H, Teramatsu T, Okamura S, Hino T, Shibata U (1981). Tracheal reconstruction using collagen-plastics composite mesh, Jinko Zoki (Art. Org.) 10 (2): 510–513

4.3g Tamura Y, Nakamura T, Mizuno H, Kato H, Shimizu Y, Teramatsu T, Hino T (1983). Polyvinyl-alcohol-silica composites and its utilization as artifical blood tubes, Jinko Zoki (Art. Org.) 12 (1): 166–169

4.4a Sakurada I, Okamura S, Inagaki H (1960). γ-Ray induced reactions of polymers, Bull Chem Text Inst Kyoto Univ 15: 145

4.4b Okamura S, Yamashita T, Higashimura T (1956) Bull Chem Soc Jpn 24 (6): 25

4.5   We have attempted to measure the amount of ^{13}C-labelled collagen polymers that are coated on the artificial organ and then transferred into a body tissue, but our efforts are not successful yet.

4.6   See also pp 54 ff of my short book, "Ko To Mure (Part and Group)", which was published to commemorate my 70th birthday (1983)

4.7a Okamura S (1990) "Seisuu No Ura To Omote (Back and Front Sides of Integers)" (K. Ei No. 167), K. K. 39 (4): 169–170

4.7b Okamura S (1991). "Kagaku Ni Asbou (Enjoy Science)", PHP Research, 73–78, 182–188

4.8a Okamura S (1989) " 'Genpatsu' To Iukoto To, 'Genpatsu No Mondai' To Iukoto No Chigai (Difference between 'Nuclear Power Generation' and 'Nuclear Power Generation Problem)" (K. Ei No. 117), K. K. 40 (2): 66–67

4.8b Okamura S (1989). "Kagaku No Shakaishi (Social History of Science and Technology)" (K. Ei No. 103), 38 (12): 588–589

4.9   Okamura S (1992). "Is depolarization for process or for goal?" (K. Ei No. 136), K. K. 41 (9): 432–433

4.10 Belloni-Coflar J, et al. (1991) Endeavor, New Series 15 (1): 2–9

4.11 Michell RD, Nebe WT, Hardam WH (1986) J Imag Sci 30: 215

4.12 Ito H, Willson CG (1983) Polym Eng Sci 23: 1019

4.13 The society of polymer science, Japan (1990) Kobunshi (High Polymers) 39 (11)

4.14 Imanishi T (1980) History of aesthetics. Tokyo Univ Press, Tokyo, p 173 "Ästehtik von unten" (1871), Fechner GT (1873) in Zur experimentalle Ästhetik

Chapter 5

Plastics and Rubber

Bengt Rånby

5.1
Developments Before 1950 – The First Generation of Plastics

The first synthetic polymer produced for commericial applications was *phenolfor-maldehyde resin* reinforced with cellulose fibers or sawdust. It was invented in 1907 by Leo Baekeland, USA, and marketed as *Bakelite* a few years later. During the early 1930s a few other synthetic polymers were developed in the USA and Germany and used as commercial plastic materials, e. g., *polystyrene, poly(vinylchloride)* (PVC) and *poly(methylmethacrylate)*(PMMA). They are vinyl polymers synthesized by free radical polymerization. Later in the 1930s and the early 1940s other types of polymers were synthesized and marketed as plastics, e. g., *unsaturated polyesters* crosslinked with styrene in Germany (1941) and reinforced with glass fibers in the USA (1942), *epoxy resins* as Araldite in Switzerland (1936) and *poly(tetrafluoroethylene)* as Teflon in the USA (1939). All these plastic materials are still produced commercially and used all over the world [5.1].

During the 1940s, polyamide from adipic acid and hexamethylene diamine was produced in the USA and marketed as *6,6-nylon* fibers. This production was based on research work by Carothers in the 1930s at Du Pont [5.2]. The same polyamide was also developed as a plastic material of high mechanical strength and high melting point (about 255 °C). Another type of polyamide was invented by Schlack in Germany who synthesized *6-nylon* from caprolactam a few years later [5.3]. This is also an excellent fiber and plastic material with a lower melting point (about 220 °C) than 6,6-nylon.

Linear polyesters were first prepared by Carothers in the 1930s, but they were all aliphatic and of too low a melting point (about 100 °C) to be of commercial interest as fiber and plastic material. A most important invention was made by Whinfield in England in 1941 who synthesized poly(ethylene terephthalate) with a melting point of 265 °C[5.4]. This polyester is today a dominant polymeric material of high strength and durability and used for fiber, film, bottles, etc. Several polyester modifications are now produced. The chemical structure of the first commercial polymers is shown in Table 5.1.

Okamura, Rånby, Ito (Eds.): Macromolecular Concept and Strategy
© Springer-Verlag Berlin · Heidelberg 1996

Table 5.1 Chemical structure of commercial polymers developed before 1950

Polymer	Chain Unit	Main Use
Polystyrene	$-CH_2-CH-$ (with phenyl ring)	Plastics Film
Poly(vinylchloride) PVC	$-CH_2-CH-$ Cl	Plastics Film
Poly(methyl-methacrylate) PMMA	$-CH_2-CH-$ CO OCH_3	Plastics Sheet Film
Unsaturated Polyester	$-O-\overset{O}{\overset{\|}{C}}-CH=CH-\overset{O}{\overset{\|}{C}}-O-$ $-O-CH_2-CH_2-O-$ $-O-\overset{O}{\overset{\|}{C}}\overset{O}{\overset{\|}{C}}-O-$ (with benzene ring)	Thermosetting Resin Crosslinked with Styrene
Epoxy Resin	$R-CH_2-\overset{O}{\overset{/\backslash}{CH}}-CH_2$ $CH_2-CH\underset{O}{\overset{}{-}}CH_2$	Thermosetting Resin Crosslinked with Amine or Amide
Poly(tetrafluorethene)	$-CF_2-CF_2-$	Plastics Surface Coating
Poly(hexamethyl- eneadipamide), b,b-Nylon	$NH-(CH_2)_6-NH-$ CO $(CH_2)_4-CO-$	Fiber Plastics
Poly(ethylene-terephthalate)	$-O-(CH_2)_2-O$ $-OC-$(benzene ring)$-CO$	Fiber, Film

5.2
Construction Plastic Materials – The Second Generation of Plastics

Polyethylene was first synthesized by Fawcett in England during the Second World War. It was first used as an advanced electrical insulation material of low electric loss in radar technology. This was low density polyethylene with low melt temperature ($\sim$ 100 °C) and branched chain structure, produced by free radical polymerization at high pressure and elevated temperature. Great discoveries were made in the 1950s by Karl Ziegler in Germany and Guilo Natta in Italy. They found that ethylene and other olefins could be polymerized to linear and stereoregular polymers using metalloorganic catalysts of titanium chloride and aluminum alkyl. This discovery dominated polymer research and development for several years. Both *high density polyethylene* of linear (unbranched) structure (melt temperature about 140 °C) and *crystalline polypropylene* of stereoregular/isotactic structure (melt temperature about 175 °C) were prepared with Ziegler-Natta catalysts and marketed as construction materials

Table 5.2 Chemical structures of construction plastic material

Polymer	Chain Unit	Main Use
Polyethylene Linear Branched	$-CH_2-CH_2-$ $-CH_2-CH-$ $-CH_2-CH_2-$	Thermoplastics Film Sheet Fiber
Polypropylene	$-CH_2-CH-$ $\quad\quad CH_3$	
Polycarbonate	$-O-Ph-O-\overset{O}{\overset{\|}{C}}-O-Ph-O-$	Thermoplastics
Polyurethane	$-R_1-NH-\overset{}{\underset{O}{\overset{\|}{C}}}-O-R_2-$	Plastics Elastomer Surface Coating
Polysulfones	$-\langle\bigcirc\rangle-\overset{O}{\underset{O}{\overset{\|}{\underset{\|}{S}}}}-\langle\bigcirc\rangle-$	Plastics

for plastics, film and fiber use. Polyethylene and polypropylene are now produced in the largest quantities of all polymeric materials.

A number of other polymeric construction materials were invented and produced during the 1950s, e. g., *polycarbonates* which are transparent plastics of very high impact strength, *polyurethanes* which are used as tough plastics, coatings and versatile foams in large scale, *polysulfones* of various types as thermoplastics of high mechanical strength and high softening temperature. These materials are called "*engineering plastics*" and have found extensive use in machine construction. They have higher mechanical strength than the first generation "*commodity plastics*" and they have softening temperatures well above 100 °C. The structure of some engineering plastics is given in Table 5.2.

5.3
Specialty Polymeric Materials – The Third Generation of Plastics

Since about 1965 a third generation of specialty polymers have been synthesized and brought to the market. They are usually of more complex chemical structure than the engineering plastics, they have very high softening temperatures (up to 300 °C) and high chemical resistance. Some of these new materials are commercial for specialized applications, e. g. *poly(phenylene sulphide)* as Ryton, *polyimides* as Kapton, *aromatic polyesters* as Ekonol, *aromatic polyamides* as Nomex and Kevlar and *fluoropolymers* as Teflon and Viton (plastics and elastomer, respectively). The chemical structures and some important uses of specialty polymers are given in Table 5.3.

Table 5.3 Speciality polymeric materials

Polymer	Chain Unit	Main Use
Poly(p-phenylenesulfide)	—⟨◯⟩—S—	Plastics (Ryton)
Polyimides	(imide ring structure with N and O groups)	Film (Kapton) Surface Coating
Aromatic Polyesters	—C(=O)—⟨◯⟩—C(=O)—O—⟨◯⟩—O—	Plastics (Ekonol)
Aromatic Polyamides	—N(H)—⟨◯⟩—C(=O)—N(H)—⟨◯⟩—C(=O)—	Fiber (Kevlar)
Poly(perfluoropropylene)	—CF$_2$—CF(CF$_3$)—	Elastomer (Viton)

5.4
Electrically Conducting Polymers

It has been known since the 1950s that certain polymers are electrical semiconductors, i. e., they have conductivities of the order 10^{-6} to 10^{-3} (ohm cm)$^{-1}$. These polymers have aromatic or conjugated chain structure where electron pairs in π orbitals are delocalized. The orbitals form electron structures which are more or less mobile along the chain in an electric field. It was discovered in 1977 that polyacetylene with conjugated chain structure couid be doped by electron transfer in various ways, e. g., chemical or electrochemical doping [5.5]. In this case doping means formation of unpaired electrons located on the chains either as excess electrons (negative doping) or as electron deficiencies ("holes", positive doping) as shown for polyacetylene in

Table 5.4 Chemical structure of electrcally conducting polymers

Polymer	Chain Unit
Polyacetylene	—CH=CH—
Polypyrrol	(pyrrole ring with N–H)
Polythiophene	(thiophene ring with S–H)
Poly(p-phenylene)	—⟨◯⟩—
Poly(p-phenylenesulfide)	—⟨◯⟩—S—
Poly(p-phenylene-vinylene)	—⟨◯⟩—CH=CH—
Polyaniline	—⟨◯⟩—N(H)—

$$R^{\oplus}$$
$$-CH=CH-(CH \overset{\ominus}{=} CH)-CH=CH- \qquad A$$

$$R^{\ominus}$$
$$-CH=CH-(CH \overset{\oplus}{=} CH)-CH=CH- \qquad B$$

Fig. 5.1a, b. Doping of poly(acetylene) chain: A negative; B positive

Fig. 5.1. In addition to polyacetylene and its derivatives, a number of other polymers with conjugated structure have been synthesized and doped to conducting polymers, e. g., *polypyrrol, polythiophene, polyaniline* and *polyphenylene*, with chain structures shown in Table 5.4. The doped polymers have conductivities from 10^0 to 10^4 (ohm cm)$^{-1}$, i. e. well into the regime of metal conductors which is 10^{-2} to 10^5 (ohm cm)$^{-1}$. Some of the conducting polymers have been developed into commercial products, e. g., in rechargeable batteries of polyaniline and as conducting membranes or fabric containing polyacetylene or polypyrrol as recently reviewed [5.6].

5.5
Liquid Crystal Polymers (LCP) and Other Specialty Polymers

As outlined in a previous chapter, the development of *liquid crystal polymers* (LCP) has its background in observations of the liquid crystal phenomenon for cholesterol esters which were made more than 100 years ago. The calculations of solution viscosity of rod-like polymer chains by Flory in 1956 initiated new interest in this area [5.7]. Flory's predicted viscosity minima for LCP in solution were observed experimentally by Kwolek and Morgan at Du Pont in 1977 [5.8]. They prepared aromatic polyamides from paraphenylenediamine and paraphenylenedicarboxylic acid containing rod-like polymer chains which could be spun from solution of an intermediate polymer concentration to strong fibers (Kevlar). The viscosity minimum is due to alignment of the rod-like polymer chains to bundles in solution which gives lower solution viscosity than free chains at lower concentrations. At higher concentrations the solution viscosity increases sharply due to interaction between the chain bundles. A viscosity minimum can also be obtained at an intermediate temperature. The spinning of Kevlar fibers is made from a solution of (poly-*p*-phenylene terephthalamide) in concentrated sulfuric acid (99. 8 %) at 70–90 °C into a coagulant bath of water at 5 °C. The polyamide chains are oriented by stretching and crystallized while the sulfuric acid diffuses out of the polymer gel. In recent years a large number of LC polymers have been prepared and processed into fibers, films and plastics of high strength and high melting temperatures. The rod-like polymer segments are called *mesogenic groups or segments.* They can be part of the main chain, connected with flexible chain segments. They can also be rod-like sidechains and give intricate crystallization phenomena.

The mesogenic segments in LC polymers contain rigid groups, e. g., aromatic esters, aromatic amides, aromatic ethers or biphenyls (Table 5.5).

For structural applications, aromatic polyester groups are important in the LCP

Table 5.5 Mesogenic groups in liquid crystal polymers

Structure	Name
	Biphenyl
	Diphenylether
	Phenylbenzoate
	Benzamide
	p-Phenylene – terephthalamide
	Hydroxy – naphthoic acid

materials, e. g., poly(hydroxybenzoic acid) (PHBA) and poly(p-phenyleneterephthalate) (PPT) with melting points as homopolymers in the region 500–600 °C. This is the temperature range where thermal decomposition is initiated. The polyester chain Ystructures are modified by introduction of other groups, e. g. p-phenyl, p-biphenyl, ring substitution, m-phenyl and naphthoic groups, which generally lower the melting point by decreasing the crystalline order of the LC polymer segments.

Introduction of flexible spacer segments between the mesogenic chain elements lowers the transition temperatures of an LC polymer to a desired range for processing without thermal decomposition. The flexible spacers may be $(CH_2)_n$ groups of a length from n = 4 to n = 10 between the aromatic polyester mesogens which brings the melting range down below 400 °C and makes processing possible without decomposition.

LC polymers with mesogenic groups as side chains are prepared by polymerization of vinyl monomers with rod-like substituents. Such monomers can be polymerized in LC state which gives new possibilities to vary and control the molecular architecture of the resulting LC polymer. A recent review of the development of LC polymers indicates the importance and the rapid growth of the field [5.9].

5.6
Elastomers

Elastomers (or rubbers) are polymeric materials with the unique property of instantaneous recovery after strain, i. e., after deformation by application of stress (rubber elasticity). The elastomers are used at temperatures above the glass temperature. This means that the polymer chains are flexible and therefore deformed when stress is applied. All elastomers used commercially are crosslinked to reduce permanent deformation by flow, i. e., the polymer chains form a coherent network which prevents the polymer chains from slipping or flowing when the rubber in bulk is deformed. In thermoplastic elastomers the crosslinks are mechanical ties of the chain ends of three-block copolymers, e. g., polystyrene end blocks on a polybutadiene middle block.

The efforts to prepare synthetic elastomers can be traced back to 1906 when attempts to polymerize isoprene were made in Germany. This monomer was selected because natural rubber used industrially since the early 1800s was known to give isoprene when pyrolyzed. At that time the polymer chain structure of natural rubber was not known. The first attempts before 1910 were directed towards preparing the cyclic dimer of isoprene, dymethylcyclooctadiene, which according to Harries was the basic compound in natural rubber [5.10]. These experiments failed of course because the dimer obtained was a liquid and showed no elastic properties. Both butadiene and isoprene were synthesized from phenol and *p*-cresol, respectively. A successful but very slow preparation of rubber-like materials from isoprene and butadiene in an aqueous emulsion with egg albumin, starch or gelatin as emulsifier was made in 1912 by Hofmann at the Bayer Co. in Germany. During the First World War the supply of natural rubber was scarce at times because of the blockade. As a substitute, methyl rubber was produced in Germany from 2,3-dimethybutadiene and used for tires, hoses, gas masks and casings for storage batteries in submarines.

The big price fluctuations of natural rubber in the 1920s and 1930s stimulated new attempts to produce synthetic rubber both in Western Europe and the USA. Copolymerization of butadiene or isoprene with styrene or methylstyrene to *Buna-S* in Germany was reported in 1933. A similar recipe was used in USA during the Second World War to produce large amounts of styrene – butadiene rubber (GRS or SBR). Another synthetic rubber developed in Germany in the 1930s was a butadiene-acrylonitrile copolymer which, under the tradename *Buna-N*, was used as an oil and gasoline resistant rubber of better aging resistance than Buna-S. It is still produced and sold as nitrile rubber. An original development in USA was the polysulfide rubber *Thiokol* which is a condensation product of 1,2 dichloroethane with sodium polysulfide. It was first used for shoesoles but later mainly utilized as rocket fuel. Thiokol has limited stability and decomposes slowly at room temperature to gaseous sulfides with unpleasant odor. These industrial developments are reviewed in two books [5.11, 5.12].

The first synthetic rubber produced in the USA was poly(2-chlorobutadiene) or poly(chloroprene) which was developed at Du Pont in 1931 and sold as *Duprene* or *Neoprene*. It is more oil and aging resistant than natural rubber and Buna-S.

Butyl rubber is based on cationic polymerization of isobutylene at low temperatures. The original homopolymer produced in Germany could not be vulcanized. Copolymerization of isobutylene with a few percent of a diene monomer, e. g., isoprene, introduced a sufficient number of double bonds in the chain for vulcanization. In this form, butyl rubber is used as a specialty rubber of very low permeability for inert gases and water vapor, e. g., for inner tubes in tires. The five types of synthetic elastomers described here, Buna-S (SBR or GRS), Buna-N(nitrile rubber), Neoprene^R (Duprene), Thiokol (polyethylenesulfide) and butyl rubber (polyisobutylene) are all produced today in large quantities. About two thirds of all elastomers used are synthetic. A more recent development is *cis* (1,4 – polybutadiene) and synthetic polyisoprene, i. e., *cis* (1,4 – polyisoprene), which both have excellent mechanical properties, i. e., low inner friction during deformation.

5.7
The World-Wide Polymer Industry

The polymer production was in its infancy in the 1940's and it is now, 50 years later, a world-wide industry of large dimension. The produced volume of plastics exceeds now the volume of iron and steel. Three industrial areas are leading this development, United States of America, Western Europe and Japan, which together haver 69 per cent of the world production of plastics but only 15 per cent of the world population. In the three areas mentioned the annual plastics consumption is approaching 100 kg per capita compared to less than 20 kg for the world average (5:13).

The main polymeric materials produced are commodity plastics while the speciality polymers giving hight performance materials are only 3 to 4 per cent of the plastics production. The high-performance materials for advanced applications, e.g. in aircraft, automobile, electronic and biomedical production, show the highest growth rate and also remarkably rapid technical development based on advanced research (5:14). The future of technology is dependent on the development of new advanced materials.

5.8
References

5.1   A general review of the early development of plastics is: Kaufman M (1962) The first century of plastics: celluloid and its sequel. Plastics and Rubber Institute, London
5.2   For a review of Carothers's work on polyamides: Morawetz H (1985) Polymers. The origins and growth of a science, New York
5.3   Schlack P (1941) US Pat 2, 241, 321
5.4   Whinfield J T (1943) Chem Ind 62: 354
5.5   Shirakawa H, Louis E J, MacDiarmid A G, Chang C H, Heeger A J (1977) J Chem Soc, Chem Commun 578
5.6   A recent review on conducting polymers by Schoch K F and Saunders H E, IEEE Spectrum, June 1992, p 52–55. See further "Conjugated polymers and related materials" (eds. Salaneck W R, Lundström I and Rånby B) (1993). Oxford Univ. Press
5.7   Flory P J (1956) Proc. Royal Society, London. A 234, 60, 73
5.8   Kwolek S L and Morgan P W (1977) Macromolecules 10, 1390
5.9   Donald A M and Windle A H (1992) Liquid crystalline polymers. Cambridge Univ. Press
5.10 An account of this early attempts to synthesize rubber is given some 30 years later by Hofmann F (1936) Chem. Z. 60, 693
5.11 Brydson J A (1978) Rubber Chemistry. Applied science publ., Barking, UK
5.12 Eirich F R (1978) Science and technology of rubber. Academic press, New York
5.13 Pöcksteiner E (1993) in *Polymers to the Year 2000 and Beyond. A Memorial Symposium for Herman F Mark* (S.M. Atlas, E.M. Pearce and F.R. Eirich, eds.), J. Polym. Sci., Polym. Symp. *75*, 137.
5.14 Yoda, N. (1993), *ibid. 75*, 125.

Chapter 6

Recent Progress of Fiber Technology and Applications
– the Creation of New Values –

NAOYA YODA

Abstract

The recent progress of fiber technology and applications is reviewed in terms of new value concept of Shingosen (New Synthetic Fibers) in Japan. Future directions of specialty products of functional fibers are discussed in terms of sensitivity and new values.

As to high performance fibers for aerospace and communication industries, the new technology and development of high-tech products such as carbon fibers for advanced composites for the transportation industry, and plastic optical fibers for the communication industry are reviewed. Finally, management strategy of the fiber industry in the age of high technology is discussed in the highly competitive borderless global economy approaching the twenty first century.

6.1
Overall Review

6.1.1
Business Trends in Japanese Fiber and Textile Industry

Industrial globalization is the current major managerial subject that must be addressed for the 21st century. The free market economy indicates that the society has entered into an age of coexistence and role-sharing relinquishing the age of confrontation and competition of the East and West. The fiber and textile industries involve two of the three essentials of human social living, namely clothing, foods and housing. Thus, fiber technology will contribute to the economic development of NIES, ASEAN and other Pan-Pacific Rim countries (APEC) by mutual interdependence and cooperation in terms of *symbiosis* of ecology in the international community.

Innovation is the driving force of economic development, and the fiber industry is no exception. In this chapter, the resent progress and trends of synthetic fiber technology and applications in the Japanese fiber and textile industries are reviewed in terms of creation of new values in technoglobalism [6.1].

The synthetic fiber industry is growing in the NIES and ASEAN nations, although stable in advanced countries such as Europe, the United States and Japan. Figure 6.1 shows the trends of world synthetic fiber production volume and demand-supply

Okamura, Rånby, Ito (Eds.): Macromolecular Concept and Strategy
© Springer-Verlag Berlin · Heidelberg 1996

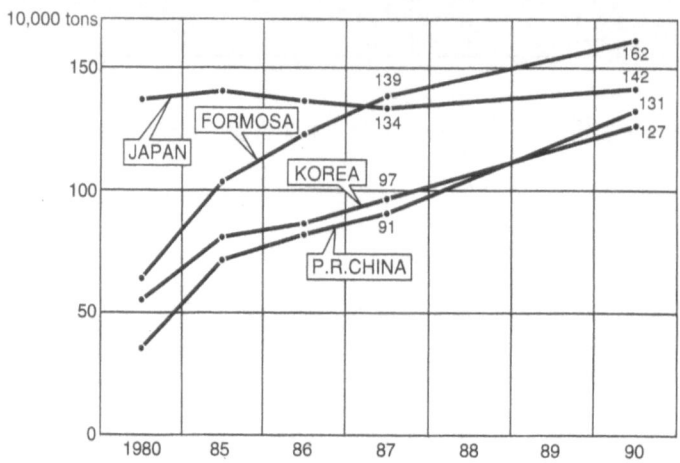

Fig. 6.1 Trends of production volume in the major Asean nation of synthetic fiber producers (Source: Textile handbook (Textile Organon 1990))

relations projected for the period of 1980–1993. The solid line shows the actual output and the world production capacity is 38–180 % higher than the worldwide demand each year. Specifically, it suggests that the world's synthetic fiber industry has been confronted with a massive surplus capacity over the past decade.

Industry in Europe, the United States and Japan, in particular, has suffered from excess capacity, while the Far East region, including South Korea, Taiwan, China and

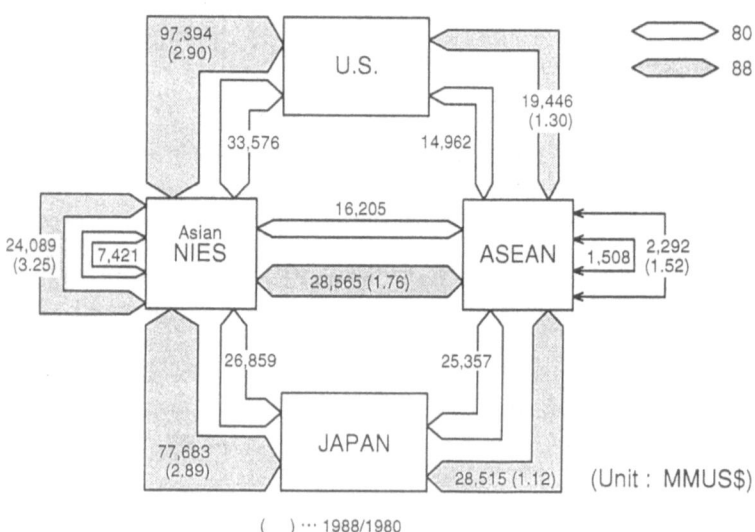

Fig. 6.2 Trade in Pacific RIM 1980, 1988

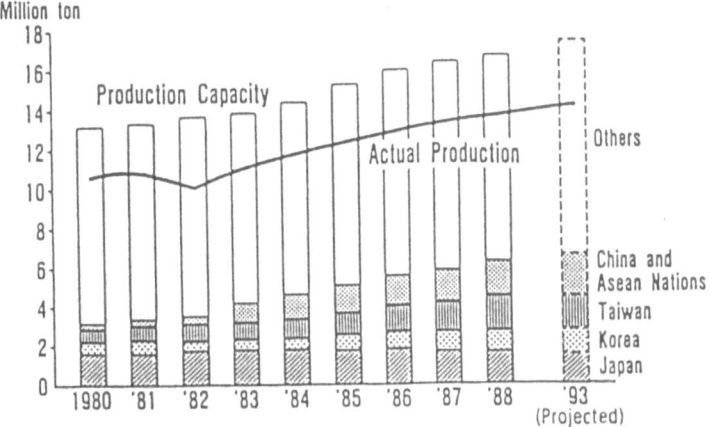

Fig. 6.3 World production of Synthetic fibers (1994)

ASEAN countries (excluding Japan) recorded output increases in the period 1980–90 and the formation of the Far East textile complex is shown in Fig. 6.2. In Fig. 6.3 the textile exports around the world are described. The solid portion of the bar chart represents exports from Hong Kong, China, Taiwan, the Republic of Korea and Japan. Of total exports, these countries account for more than one-third, which is outstanding. Starting with Hong Kong's U. S. $ 18 billion, South Korea, China and Taiwan are all big exports, earning more than $ 10 billion each. When the trade relations of Asian NIES and ASEAN nations with U.S. and Japan are compared between 1980 and 1988, as shown in Fig. 6.4, trading volume is seen to have increased almost three time between NIES and both Japan and U.S. whereas it increased more than 76 % between ASEAN and NIES countries.

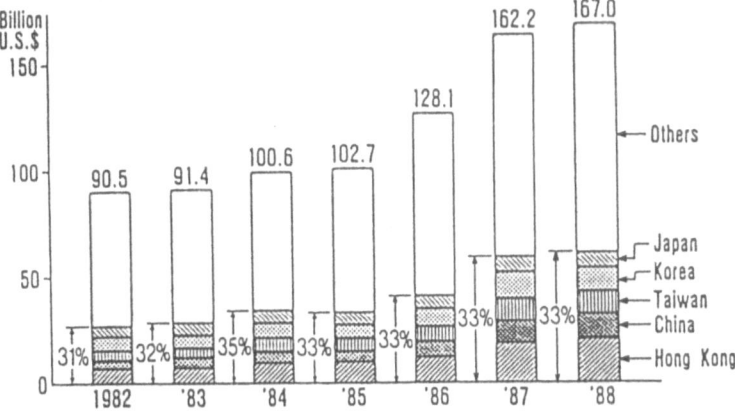

Fig. 6.4 Textile export sales from Far East bloc (1990)

Therefore, the pan-Pacific region acts as the world's largest trade center. The region is the world largest production center on the one hand, and forms the leading consuming area on the other. The author personally calls it the Far East Textile Complex, and believes that, without sound development of this textile complex in the years to come, we can never expect stable growth in world textile production. Therefore, Hong Kong and Japan above all are expected to play key roles, with the former as the largest textile trading country and the latter as an advanced textile producing country.

6.1.2
The Japanese Synthetic Fiber Industry Entering a Period of Maturity

The size of the fiber industry in Japan has been compared with five other major manufacturing business sectors in terms of number of employees, sales volume and gross added value in 1989 as summarized in Table 6.1. With eleven million employees in all fields, the textiles sector employs about 560,000 (5 % of total number), whereas shipment amounts to 7.6 trillion yen (2.7 %) and gross added value is 3 trillion yen (3 %). The trends of Japan's synthetic fiber output records are shown in Fig. 6.5. Production had increased until the early 1970s. After the Smithsonian Agreement in 1971 and the first oil crisis in 1973, however, Japanese industry came to face excess capacity, which became conspicuous as a result of sluggish local demand and narrowing export markets. Plunged into desperate business conditions, the industry had no choice but to resort to government support in its efforts to dispose of excess capacity and to dissolve unprofitable exports. By the collapse of the bubble economy in 1991, however, the fiber industry has been in recession and gradual recovery is expected in 1995, followed by sustainable growth in late 1990. Total production in Japan is about 1.4 million tons per year, of which polyester fiber represents 36 %, acrylics 29 % and nylon 21 %.

Table 6.1 Chemical Industry in Japan

Industries	Employees (10^3 person)	Shipment (Tri. Yen)	Gross Value Added (Tri. Yen)
Chemical	390	20.5	10.1
Electric / Electronics	1,900	46.8	17.2
Automobiles	890	37.4	9.9
Steel / Metals	500	22.0	7.4
Textiles	560	7.6	3.0
All	10,900	274.0	102.7

Source: MITI (1989)

Table **6.2** Development of Synthetic Fibers through the Phases

Phase	I	II	II
Features	Life's Necessities	Added-features for Living	Cultural Life Style
Role	Replacement and Supplement to Natural Fibers	The Right Fiber for the Task Fiber Combinations	High-Technology High-Desirability High-Quality Advanced Material
Functions	• Strength • Durability • Cost-Conscious	• Wash & Wear • Anti-Piling • Soil-Release • Stain-Resistant • Elasticity • Anti-Static	• Fine Textures • Sensuality • Versatility • Product Individuality • Fashion Conscious • Comfort • High Value

6.1.3
Prospects for the Synthetic Fiber Industry

Assuming that the development of synthetic fiber production processes can be divided into first phase as shown in Table 6.2, followed by second and third phase, what lies ahead is the third phase. It is believed that important objectives for the industry are to understand sophisticated, diverse consumer demands and to establish a quick response and efficient supply system to meet actual demand better. Market needs will be for a wide variety of products, each in small quantities. Thus, we must become a leading industry in the age of changing lifestyles by creating a new sense of values, and steering the consumer towards better lifestyles. Keys to solving the problem are innovation of high-technology, sensibility and high quality of specialty products. Develement phases consist of three stages of features, roles and functions as summarized in Table 6.2. As long as we can cope successfully with the changing management environment, we can survive with international competition by creating new markets.

6.2
Technological Innovation of New Fibers (Shingosen)

6.2.1
Technological Developments of New Synthetic Fibers

Technological developments in the textile industry have been very substantial, particularly over the last three decades. In this chapter a quantitative analyses is carried out and the impact of technology in fiber industry in Japan is discussed concerning technological developments since 1967 on machine and labor productivity, quality,

Table 6.3 Key Technologies of Synthetic Fibers

1. Molecular Design
2. Delicate and Sophisticated Fabricating Technology
3. Morphology Control
4. Conjugation / Hybrids / Composites
5. Characterization / Higher-Function-Added Products
6. Highly Reliable Production Technology

costs, and services. Great as these improvements have been, have they succeeded on making the textile industry of developed high-wage countries competitive, from a cost point of view, with the textile industries of lower-wage countries of South-East Asia? In order to answer this question, a comparison is made of producing a basic apparel product, a cotton/polyester-fiber dress shirt, in the United States and in South-East Asia. The results show that, for all practical purposes, a state-of-the-art mill in the U.S.A. can be competitive with a South-east Asian mill in producing the fabric for the dress shirt but that the advantage of the low-wage country in making-up the shirt is still sufficient to make the final product, i. e., the dress shirt, competitive from a pure-cost point of view. It is also pointed out that this cost advantage can be offset to a substantial degree by improved services.

The textile and apparel industry is made up of several sub-sectors with very different characteristics, ranging from synthetic-fiber production to nonwoven-fabric manufacture, and it would take a whole volume to cover the impact of technology on all these sub-sectors. In this chapter, we shall confine ourselves to the sectors that represent 75 % of total textile consumption, namely, the yarns and fabrics that are used in household textiles and apparel.

Technology has had an impact on machine productivity, labor productivity, quality, costs, services, management flexibility and response time. The progress of textile technology over the last two decades has had a considerable impact on product cost, quality, flexibility, and capital requirements, and these are all major factors in determining the overall competitiveness of a textile enterprise. To try to measure these improvements, we shall compare the situation today with that in 1967. (Table 6.3)

6.2.2
Impact of Technology on Labor Productivity

If we pursue the example used in the comparison of machine productivity, we see that the impact on labor productivity has also been substantial. In this specific example, the labor productivity is expressed as kilograms of yarn produced per operative-hour in all processes of the spinning mill, from raw material opening to winding inclusive, and including direct, indirect, and service workers. The productivity labor cost of the better U.S. mills is also considered. In 1990, with the same technology as in the U.S.A., the labor productivity was up to 47.7 kg/operative-hour, or more than eleven times the level of 1967, while the hourly wages have increased 9.5 times, resulting in a labor cost per kg of 6.4 cents, which is 18 % less than in 1967 and 41 % of the unit labor cost achieved in the U.S.A.

The improvement in labor productivity in the U.S.A. has been growing at a lower rate in both the U.S.A. and Hong Kong than the houry labor costs, resulting in a higher unit labor cost in both countries in 1990 than in 1967. The Hong Kong unit labor cost in weaving, which was only 41 % of the U. S. unit labor cost in 1967, has gone up to 55 %. The labor-cost content at each stage of manufacture of a dress shirt shows that, of the total labor-cost difference of $ 1.34, $ 1.15 or 85 % of the total is in the making-up process.

Unfortunately, true Quick Response systems are rare in the textile / apparel chain. To have one's goods bar-coded does not by itself imply that a Quick Response system is in place and functioning efficiently. Technology today is not only concerned with production machines and equipment; it is also concerned with the improvement of managerial skills, and we fell strongly that the need to be successful in implementing the high technology managerial skills that the high technology plants require will be the most important challenge facing textile and apparel firms in the 1990s.

6.2.3
Recent Trends in Development and Application of High Performance and Functional Fibers

To live better, a nation must produce well. The modernization of fiber production technology is the key factor to increase productivity and to enhance the productive performance of the fiber industry. The Japanese fiber industries have been expanded, their business markedly centering around synthetic fibers since 1945, i. e., after World War II.

Since the first oil crises in 1965, the growth rate has dropped off sharply through a variety of economic restrictions. The restructuring of the Japanese fiber industry was initiated by the rapid increase of naphtha prices and the appreciation of yen since the

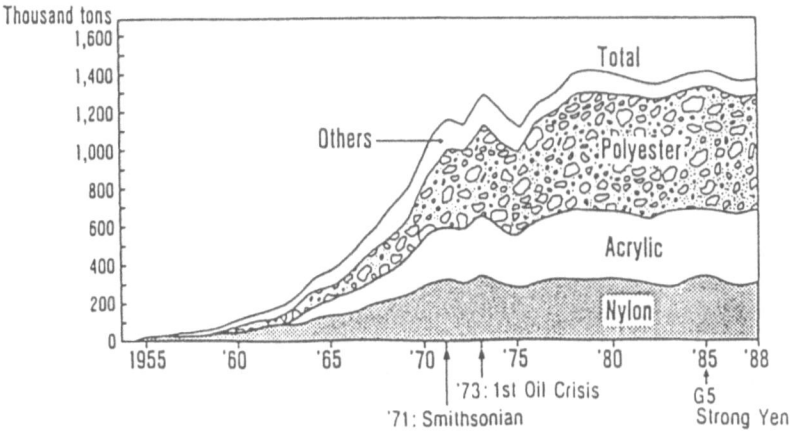

Fig. 6.5 Synthetic fiber production in Japan

Table 6.4 Key Technologies for Improvement in Hand Properties (Sensitivity) of Synthetic Fibers and Fabrics

Process	Key Technology
Melt spinning	• Microfiber or fine denier $(0.7 \sim 0.1^d)$ • Non-circuler cross section • Variation of spinning speed • Spinning of modified polymer (copolymer, polymer blend, etc.) • Conjugated yarn
Texturing	• Thick and thin yarn • Random heat set • False twisting (with drawing) • Air texturing • Combined yarn of filaments and stable fibers
After treatment and finish	• Caustic reducing (polyester) • Treatment for bulkiness • Creping • Raising • Creasing • Treatment under plasma

oil crisis in 1973. A crucial step was taken in industrial restructuring to recover from the oil shock, that materialized as structural improvements, transformation to knowledge-intensive business, introduction of high value-added products with sophisticated technology, differentiation of unique specialty products and highly functional products. Moreover, the management style was drastically changed to the customer-oriented approach, responding to market needs. Toray's new products developed by R&D in the period of 1950–1990, and the growth of fiber business are illustrated in Fig. 6.5. Among those specialty polymers, specialty fibers – so called Shingosen, – new synthetic fibers in Japan, attracted special attention for value-added products having such fiber characteristics as linear fiber-forming, isotacticity and isomorphous properties. Furthermore, the necessity for new fiber technology involving high performance fabrication processes to form fiber structures utilizing the unique, property of large surface fibers was stressed. Six key technologies of synthetic fibers are summarised in Table 6.3. Key technologies for improvement of sensitivity of fibers and fabrics are shown in Table 6.4.

In Table 6.5, the example of new fibers of highly functional polymers is shown in terms of the functionality and characteristic property of each fiber. Advanced technology development by the combination of new materials is closely related to the link between market needs and seeds derived from research and development efforts of scientists. The combination of several innovations of technology is necessary to develop new products that meet with market needs. In this case, the successful products of high value-added sophisticated technology are obtained only by the joint effort of interdisciplinary interactions among suppliers and converters vertical relations.

Table 6.5 Classification, Main Functions, and Application of Functional Fibers

Classification	Example of functions	Application
Physical and / or morphological properties with high performance	High strength / high toughness / high modulus / high elasticy labrasion resistance / low abrasion / anti-fatique / light weight / ultra-micro de ier / good processability with surface	Aeronautics / energy saving / high performance fibers
Electricity electronics	Electrical insulation / dielectricity / electric conductivity / piezo-electricity / pyroelectricity / super-conductivity / antistaticity / information memory	Insulating materials / dielectrical material / insulation of electric cord / conducting fiber / exotherMic sheet / piezoelectric polymer / information processing / energy saving
Photodynamics	Light fastness / weather resistance / light absorbability / light refractivity / light interference / light permeability / organic semiconductor / photochromism / information memory / light transferability / light selective permeability / photoelectric changeability / radiation resistivity / radiation absorbability / radiation reflectivity / electromagnetic wave screening	Optical fiber / light refraction / organic photo-conductor / X-ray absorbant or scattering material / radiation shield fiber / energy saving
Acoustic oscillation	Sound absorbability / noise insulation / vibration controllability / vibration screening / shock absorbability	Sound absorbing material / noise insulating material / vibration controlling material / vibration screening material
Magnetism	Magnetic induction / anti-magnetism / magnetic insulation / magnetic screening / information memory	Magnetic filament / magnetic screening sheet information processing / aeronautic and universal field
Heat	Heat insulating / heat conductivity / thermosensitivity / pyroelectricity / heat resistance / thermoelectricity / heat regenerativity / resistance to a low temperature / fire retardancy / flame proofing / thermochromism	Heat insulating material / thermal insulating material / cooling material / plastic thermistor / pyroelectric material / thermochromic material / energy saving

6.2.4
Improvement of Surface Properties of Polyester

The textile industry in Japan creates specialty and proprietary products based on two areas of expertise: its high-speed, precision technology and hybrid and mixing technologies. Key technology of synthetic fibers usually consists of several major categories. There are two major approaches – micro-level approaches of molecular design, polymerization and morphology control with polymer alloy concept. Moreover, there

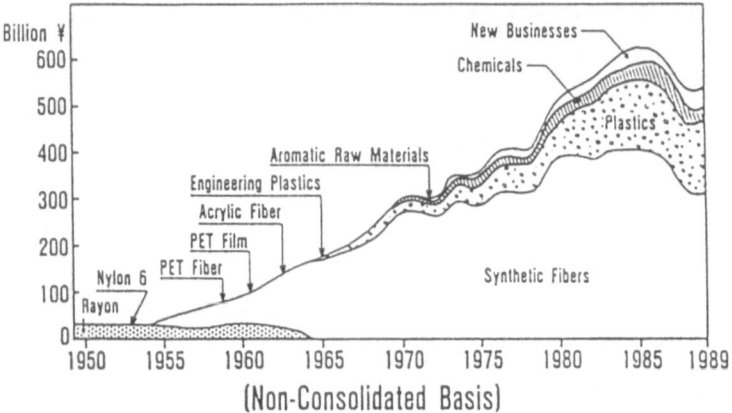

Fig. 6.6 Toray's sales und products

are macro-level approaches such as spinning process development, delicate and sophisticated fabrication and highly reliable production technologies.

The trends in fiber technology in terms of market requirement from low-tech to high-tech versus low-value to high-value added fiber products are summarized in Fig.6.7.

Increases in productivity in the textile industry are largely based on increases in production speed. At the fiber-production stage, the speed at which synthetic fibers are spun has increased from 1000 m/min in the 1960s to 7000 m/min or more. The important point here is that quality factors such as fiber strength and evenness of liner density and length are realized precisely as planned. This is particularly important in realizing a corresponding increase in production speed at the next stage, where the yarns are woven into textiles [6.2, 6.3]. As for hybrid and mixing technologies, many examples can be given at the fiber stage. One example is new synthetic fibers which are made of two polymer components spun at once. A minute difference in heat shrinkage between the two components results in a new texture.

Fibers constructed much like a pencil, with a core of carbon fiber and an outer covering of polyester fiber, are another example. Because this type of fiber can be used to earth static electricity, it is now indispensable for clean-room wear at semiconductor production facilities.

Increases in textile-production speed have been as impressive. The speed at which the weft is woven in has leapt from 150 times/min in the days of the flying shuttle to 400 with the rapier loom and 600–700 times/min or as high as 1500 times/min on the water-jet loom. These increases in speed were made possible by adopting totally new and innovative ideas for the mechanism by which the weft thread travels back and forth.

These increases in production speed mean that, if looms should have to be stopped because of thread breakage, the time lost is much more costly. Japan still exports some fibers that are the same grade as those produced in newly industrialized economies. However, the Japanese fibers have a high reputation for quality and can be relied on not to cause expensive delays due to breakage.

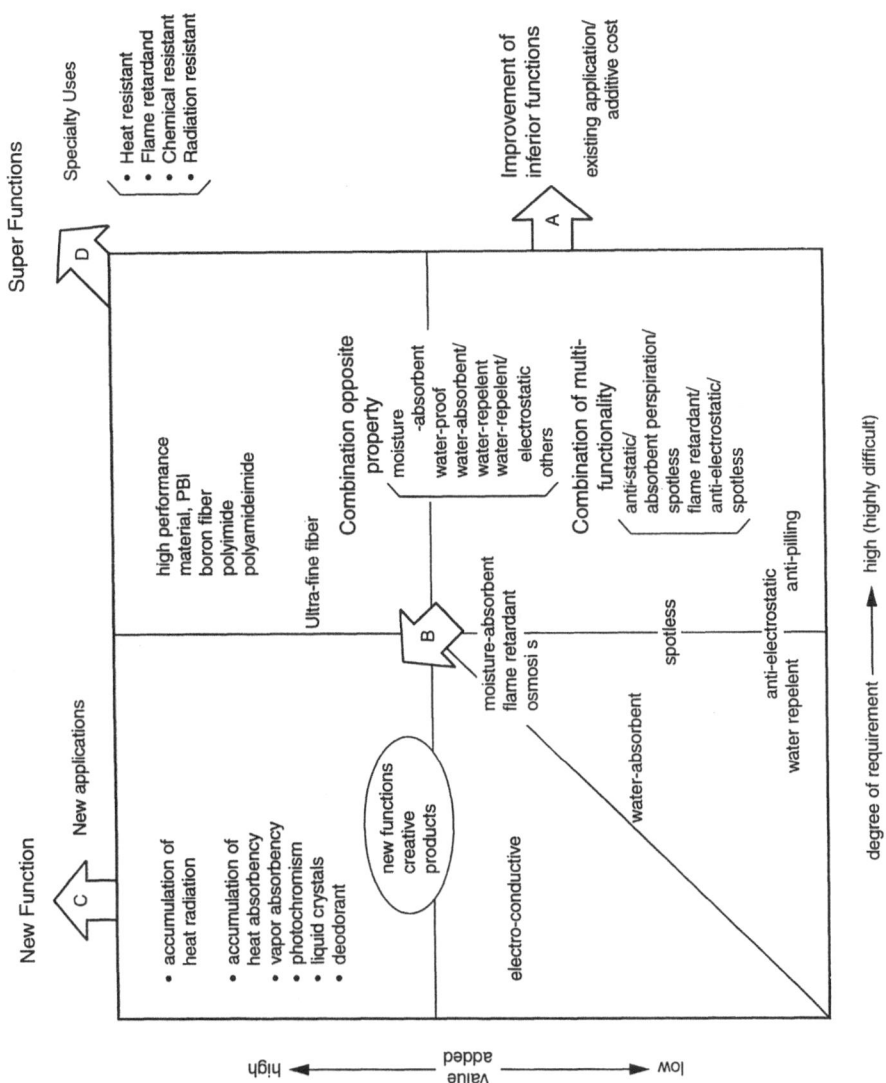

Fig. 6.7 The trends of fiber technology in terms of market requirement from low-tech to high-tech versus low to high value-added products

Table 6.6 Process of organic material and examples of speciality products

Process flow	Raw material	Material	Primary processing	Secondary processing	High value-added Processing or Construction
Change of material example of technologies	*Chemical change* Petrochemistry Organic-synthetic chemistry Fermentation chemistry Pulp manufacturing and wood chemistry	*Chemical change* Polymerization Polymer modification	*Physical change* Spinning (extrusion) Spinning Film or sheet molding Nonwoven fabrics Plastic molding Paper making	*Morphological change* Knitting, woven Dyeing Chemical modification and finishing Versatile processing	Garment manufacture Artificial organ Utilised system of separation members Fresh water manufacture from sea water, Concentration of fruit juice, Refinement of proteins, eg, enzyme Effluent waste treatment Recovery of organic materials from organic wastes Utilized system of Magnetic materials Printing materials Imaging materials Noise insulating materials, Heat insulating materials
Technological characteristics	Mass balance and energy balance	Mass balance and energy balance	Variety of control of variousness	Variety of precise control control of sensation	Control of sensation, realization of highly efficient functions in a total system

Process flow	Raw material	Material	Primary processing	Secondary processing	High value-added Processing or Construction
Sociality products		New functional polymer semi-conductivity • photo-crosslinking Ion-exchangability Selective filtration High performance polymer Heat resistance High modulus New engineering plastics High performance fiber	Micro fiber Hollow fiber Conjugated yarn Gas-barrier membrane Antithrombogenic membrane	Fabrics of natural-fiber-like synthetics • silk-like • cotton-lie • linen-like Textiles with good drape Antistatic fabrics Ionic dyable polyester	New products for new industrial Applications

Source: TBR (1994)

Key technologies for increasing sensitivities, i. e., hand properties for new synthetic fibers (Shingosen) is described in Table 6.5. There are three major production processes, namely melt spinning, texturing, and after treatment and finishes. Furthermore, there are important key technologies recently developed for commercial applications in Japan as summarized in Table 6.5. Functional fibers are classified as electrical properties, photodynamism, acoustic oscillation, magnetism and heat properties and examples of applications are shown in Table 6.6. The new process of organic material and example of specialty products are described in Table 6.6.

The recent progress of new fiber technology for value-added functions by new processes is summarized in Table 6.7. For adding functional characteristics to the fiber, the precise control of the weaving process is a vital factor in producing the desired characteristics in the finished product. The functions required for clothing are summarized according to the demanded properties and related typical functions [6.4]. Hybrid mixes at the textile level are also being pursued, such as using continuous-filament yarn for the warp and spun yarn for the weft. Such weaving techniques are expected to produce new textures with a unique touch and drapability that were not produced before [6.5, 6.6].

Experimental tests on texture is on-going in the area of knitted products as well. Examples include knitted fabrics using synthetic fibers on the surface and cotton inside, or using yarns of different thicknesses. In the latter example, the side facing the wearer may be knitted by using thick yarn and the outside knitted by using thin yarn.

This knitted fabric will absorb moisture on the inside and evaporate it quickly on the surface (because of the greater amount of yarn surface per unit surface of product), resulting in greater comfort.

Other examples of hydbrid mixing include the production of yarn spun with a blend of two or three different fibers chosen according to the end-use, and Spandex yarn wound with extremely fine nylon yarn for use in support-type stockings. The use of shape-memory alloys for the underwear in brassieres is another hybrid that gives a clean silhouette to, women's outerwear without sacrificing comfort.

The area of clothing has been increased in the speed of sewing machines but also a greater demand for clothes that fit well and satisfy the diverse tastes of consumers.

One system that is gaining popularity is the „ready-made order", a system that combines the silhouettes of ready-made suits with a wide choice of fabric color and design. By using this system, it is becoming possible to supply right-handed people with suits that have a right sleeve that is from 0.5 to 1.0 cm longer and vice versa for left-handed customers.

Table 6.7 Fiber technology for value-added functions by new process

Item		Polymer	Melt-spinning	Fiber (Filament / spun fiber)	Weaving / knitting	After treatment of fabric
New process		copolymerization blend	conjugate shape of cross section micro structure	twisting crimping chemical treatment physical treatment	Mixed weaving mixed knitting textile weave density	chemical treatment physical treatment coating laminating
physiological comfortability	moisture absorbability (or hygroscopicity)	hydrophiric polymer		grafting	mixed with cellulosic fiber	grafting
	sweat absorbability (water absorbability)	Multi-pore (structure)	fine denier, non circuler cross section	mixed with cellulosic fiber grafting	mixed with cellulosic fiber multi-layer structure (-phase)	surface polymerization coating of polymer
	moisture penetration water proofing		micro fiber divisible fiber		high density fabrics	complex or laminate of multi fine pore film
	antistaticity	antistatic polymer				antistatic treatment
	electric conductivity	electric conducting polymer	conjugate	coating	mixed with electric conductive fibers	metalized
	thermal insulation		fine denier, foaming		multi-layer structure	vacuum evaporation of Al
	bacterial inhibition		anti-microbial agent	adsorption of Cu		anti-microbial treatment

Table 6.7 continued

Item		Polymer	Melt-spinning	Fiber (Filament / spun fiber)	Weaving / knitting	After treatment of fabric
movement	stretch	elastic polymer	conjugate	crimping and (texturing)	textile weave	stretch treatment
safety durability	flame retardant	flame-retardant polymer		fire-retardant treatment		fire retardant treatment
	melting resistant	heat-resistant polymer				melting resistant treatment
	anti soiling	hollow fiber with non circular cross section	antisoilding agent			antisoiling treatment
	tenacity	high degrees of polymerization	drawing heat-treatment			
	color fastness	ionic dyable monomer	high degree of crystallization			selection of dye

Source: TBR (1994)

6.3
New Synthetic Fibers for Apparel Applications

6.3.1
Review of Development of the Shingosen (New Synthetic Fibers)

The development of the Shingosen is reviewed in some detail, Shingosen providing a new environment of materials where no natural fiber can satisfy fashion designers. At first, the R&D was directed to improve the physical properties of polyester fiber. Silk was always an ultimate goal for synthetic fibers, and in the next step polyester fiber with a triangular cross-section was produced by mimicking silk. Since the modified cross-section yarn alone was found insufficient to mimic silk, the fabric structure was modified to resemble silk fabric by caustisizing.

Due to the excellent properties, polyester fabric has been widely used in clothing over a long period of time. In addition to preserving such essential characteristics as "wash & wear", "wrinkle free" and "work saving", a major issue in polyester fabric development has been an attempt to make such aesthetic qualities as the feel, luster and coloring, as well as such functions as sweat absorbency closer to those of natural fibers. Regarding aesthetic qualities, products featuring the special aesthetic characteristics of so-called "new synthetic fabrics", i. e., principally polyester are currently the subject of great attention [6.7].

In terms of function as well, many new products are being developed which make excellent use of the special characteristics of polyester. Various functions are demanded of the clothes we wear, bedclothes, etc. in accord with the purpose for which they are used. Table 6.8 shows typical functions corresponding to certain common demands.

For example, functions required for physiological comfort include sweat absorbency and warmth, while functions required for safety include flame retardance, heat resistance and chemical resistance, etc. No matter whether natural fabrics or synthetic fabrics, it is not easy to access these functions using just fibrous material, and a variety of processing technologies should be developed to afford desired properties. The improvement of the surface properties of polyester began with the improvement of such shortcomings relative to natural fabrics as a lack of sweat absorbency, generation of static electricity, and easy soiling. The recent trends of improvement of surface

Table 6.8 Functions required of clothing

Demanded Properties	Typical Functions Required
Physiological comfort	Sweat absorbency, Fast-drying, Warmth, Windproof, Moisture permeability, Deodorizing, Water repellency, Anti-static, Waterproof, Soil release
Safety / comfort	Durability, Conductivity, Fire / heat resistance, Chemical resistance, Radiationproof, Electromagnetic interference shielding, Anti-bacterial / anti-fungal, Soil release, Anti-irritation to skin, Nontoxic
Movability	Stretchability
Other	Anti-static, Conductivity, Lint-free

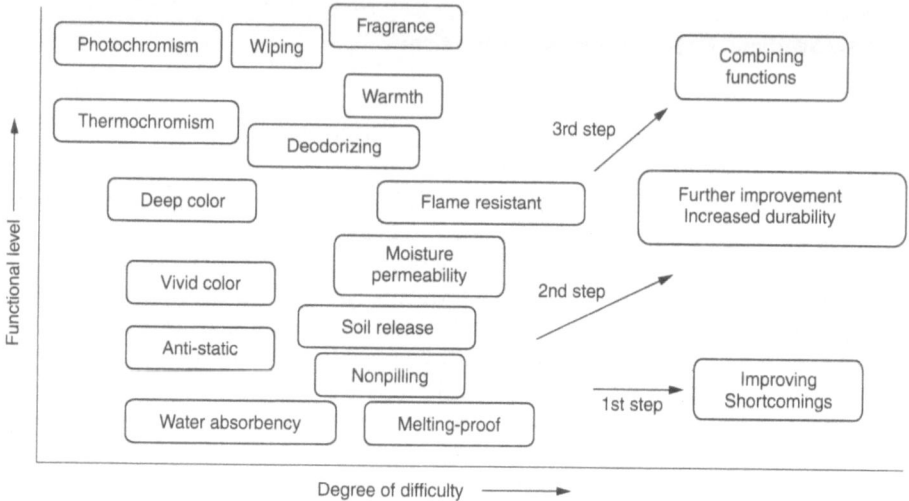

Fig. 6.8 Relationship between degree of difficulty versus functional level of fiber technology
∅ : large
∅ : small

Fig. 6.9 Structure of water repellency

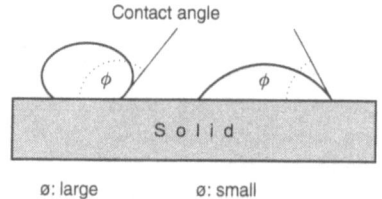

properties of polyester to give specific functions to fabrics are summarized in Fig. 6.8, The first step is the improvement of inferior properties. The second step is to pursue a variety of new functions to improve fiber durability. The third step involves combining functions to make a multifunctional fabrics (for example, combining an anti-static function with sweat absorbency, waterproof, or moisture permeability.

In Fig. 6.9, the structural model of water propellency is described in terms of contact angle φ, either large on the left hand side or small on the right hand side of the model.

These functions were obtained in the past by means of dyeing and finishing processes, which involves chemical or physical treatment of the surface of polyester fabric so as to improve its properties. With an atempt to enhance functional properties to polyester, technological innovations have been applied to the other polyester fabric manufacturing processes. It involves polymerization as well as yarn spinning and yarn texturing process. Attempts are also being made to improve polyester functions by combining the technologies used in each process.

The design concept of Shingosen is shown in the diagram in Fig. 6.10. Biomimetics suggested what should be done next, and thus special emphasis was made on the particular characteristic to surpass silk in at least one respect. There are four areas of

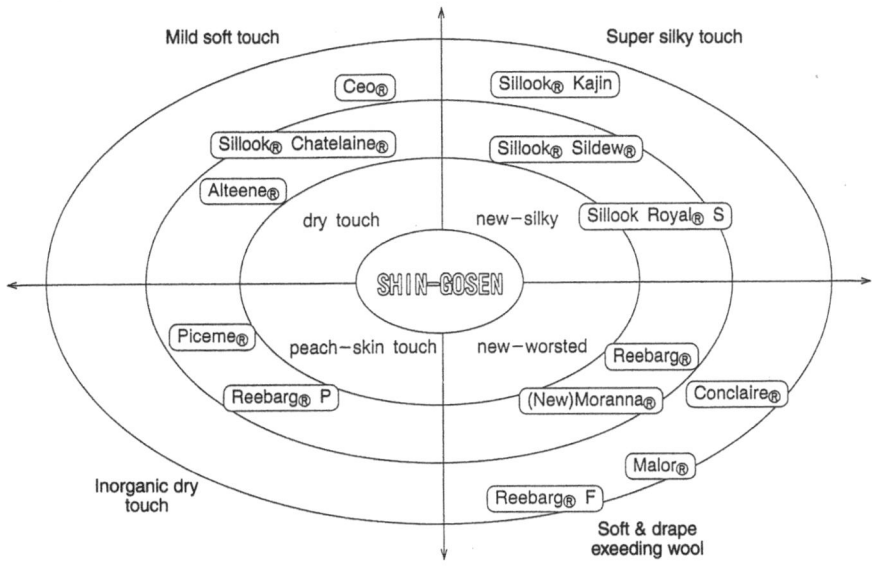

Fig. 6.10 Toray's new textile fabric of synthetic fibers (Shin-Gosen) (1994)

Table 6.9 Water Contact-Angle of Polymers

Polymer Material	Contact Angle (degree)
Poly-tetra-fluoroethylene	108 °
Paraffin	105–106
Nylon	70
Polyester	81

Table 6.10 Diversification of silk-like fibers and fabrics

Product Concept	Example (maker) of product (manufacturer)
Degummed silk like	SHANRUBY (Teijin) CYNTHIA SC (Kuraray)
Spun silk like	LUVENA (Unitika) RENNA, CRISETA (Mile) SABLINA (Toyobo)
Raw silk like	MICHELROULLAN (Toyobo) Tepla (Teijin)
Wild silk like / tussah silk like	SILART®, SILDORM®, MIXEL® X, FIVEY® (Teijin)
Super-silk like	MIXEL VII (Teijin)
Tie silk like	FESMY (Toyobo)
Hard spun silk like	SHARAIL (Teijin)
(Crepe)	SILSOIE (Teijin)
Silky wool like	MIXEL VIII (Teijin) Silmie Lhoraine (Unitika) TWISTEL (Mile)
Silky rayon like	TEPLA / Teijin)
Silky linen like	RAMIYON (Toyobo)

new design concepts of sensibility for new textile fabrics of Shingosen, namely, 1) super-silky touch, 2) soft and drape touch exceeding wool, 3) inorganic dry touch and 4) mild soft touch. There are new products recently developed by Toray in the commercial market. Recent trends in diversification of design concepts of sensibility by the major Japanese fiber suppliers are reviewed and summarized in Table 6.10.

6.3.2
Success Stories of Toray's New Shingosen Products

Diversification of silk-like fibers and fabrics are particularly successful for those makers and the product lines are shown in Table 6.10. New fiber technology of new material of mixed yarn used by the Japanese fiber producers is summarized in Tables 6.11 and 6.12. Typical examples of Toray's new Shingosen products are described below.

(1) Moisture-permeable and waterproof fabric: "ENTRANT"

The product called ENTRANT is a functional fabric which is moisture-permeable while being waterproof. It has passed the severe environmental test of behavioral assessment made under extreme conditions produced in TECHNORAMA, a weather simulation laboratory at Toray. Large amounts of ENTRANT are bought by converters for manufacturing sportswears.

(2) Silky fibers: "SILLOOK"

This product is a typical example of a highly sensuous material. SILLOOK is

Table 6.11 Polymer-modified fibers with new characteristics

Fundamental technology	Objective	Polymer	Example of Products (Manufacturer)
Co-polymerization	Basic (cationic) dyable under normal pressure	PET	LUMILET (Toray) CALAFINE (Toyobo) MELTOPE (Teijin) FINA-LON (Kuraray)
	Dispers-dyable under normal pressure	PET	BISERL (Kanebo)
	Basic (cationic) dyable (under high pressure)	Acr.	CIELO
Addition of particles	Improvement of vivid and / or deep color	PET	SN2000 (Kuraray)
	Improvement of luster	PET	CRISPEL II (Toyobo)
	New hands (dry hands)	PET	(Teijin) ALTEENE (Toray)
Conjugated melt-spun	New hands by special shape of cross section	PET	DEFORL SILLOOK ROYAL (Toray)
	Ultra-micro fiber	N	
	Stretch	PET	NEW-SRILON (Kanebo) ARTLON K (Kuraray)

Source: Toray Industries (1994)

Table 6.12 New material of mixed yarn of synthetic fibers

New technology	Example
Combined yarn of differentially shrinkable filament	SILLOOK II, IV, IVV, SILBONY, APEROLA (Toray) MIXEL III, VII (Teijin) SOANIE, SILMIE-5 (Unitika) JUNESOWAIE (Asahi Chem.) SOLRIAN, FRANSOA, TWISTEL (Mitsubishi Rayon) DELFI-NO, CRISPEL II, CRIMTOPIA, RIVIERA (Toyobo)
Multi-layer structure by false twisting of different yearn	MILPA, SHARAIL (Teijin), MORANNA, MENIMOA, TASPA, LAMBLEY (Toray) SANTOS, RAMIYON, LUCENTE (Mitsubishi Rayon)
Mixed yarn of different fibers	JOHNBELL (Kanebo) LITELAS (Toray) MIXELAT (Teijin) MISTI (Kuraray) TASTEM, SILM, PUWRUSHI (Asai)
Combined yarn of filaments and spun fibers	MANERD, BUENOCEL (Toyobo) SANOI (Toray) SAPOL (Dia Fibers)

reputed for its "more silk-like than natural silk" appearance, and, in 1991, SILLOOK celebrated its 30th anniversary on the market. After SILLOOK I, II and III, the latest SILLOOK ROYAL S has finally reached the stage where it outpaces natural silk. Fig. 6.11 shows fiber cross sections of different parts of the yarn.

(3) Ultra-fine fiber: "TORAYSEE"

A successful example of challenging the micro-fiber is "ECSAINE" developed by Toray. An example of ultra-fine denier fiber structure is shown in the microscopic cross-sections in Fig. 6.12. The product is also called ULTRA-SUEDE in the U.S. A recent successful application of this technology was a wiping cloth. It was introduced

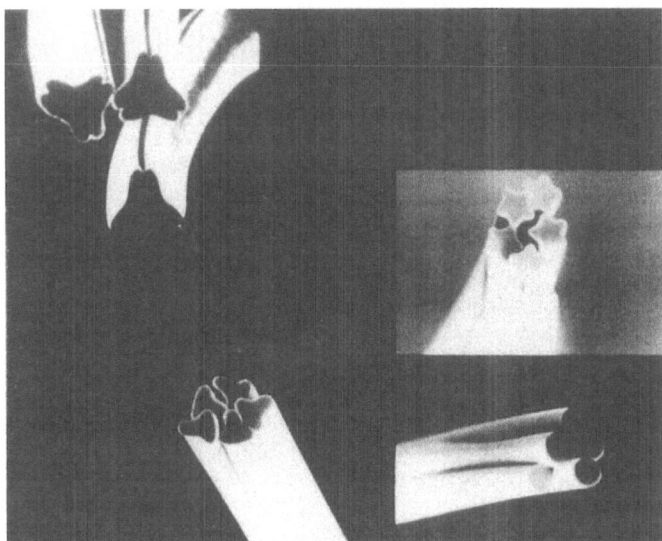

Fig. 6.11 Silky fibers - sillook[8] / silkijoy[3]
(Source: Toray Industries, Inc.)

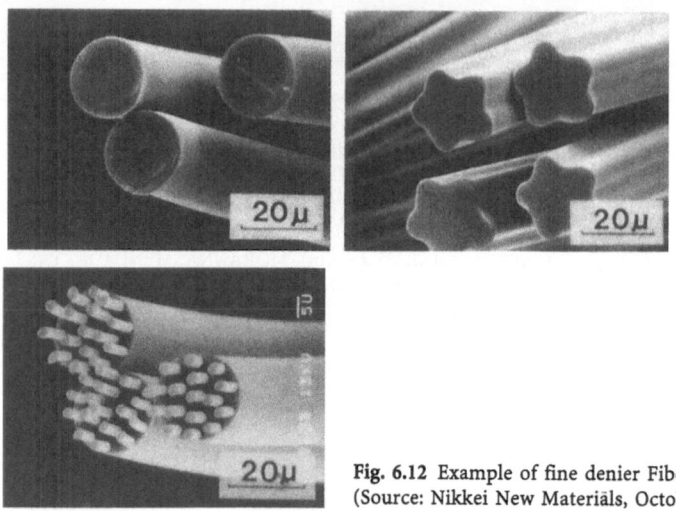

Fig. 6.12 Example of fine denier Fiber (ultra-fine Fiber)
(Source: Nikkei New Materials, October 17, 1988)

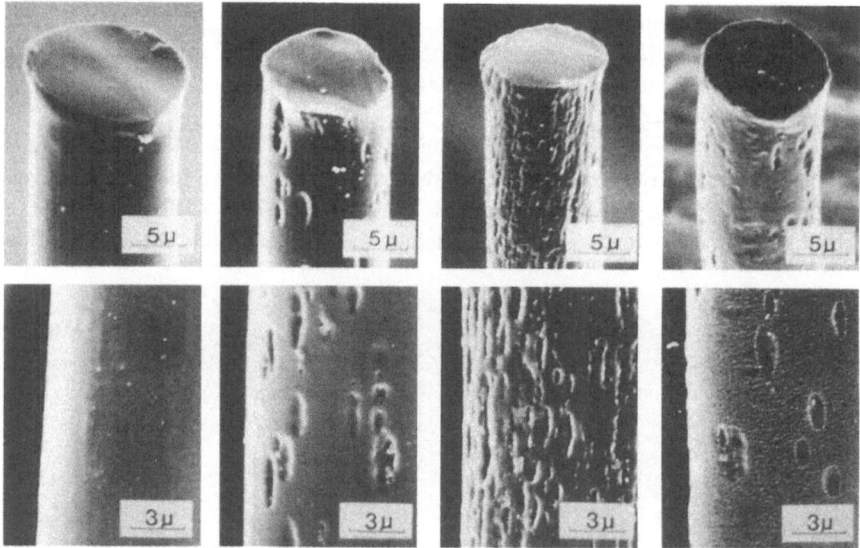

Fig. 6.13 Fine textured filament fabric – Reebarg[8]
(Source: Toray Industries Inc.)

on the market as a lens cleaner under the trade name of "TORAYSEE", and became a big hit. It can wipe off oily microcontamination amazingly well and is applicable to lens cleaning, polishing jewellery and precious metals. (Table 6.12).

(4) Fine textured filament fabrics: "REEBARG"

Today, in an age of discriminating consumers, the market is seeking sophisticated, tasteful fabrics. REEBARG is a polyester filament fabric. It has a fresh feel and appearance which are normally found in neither conventional polyester fabrics nor natural fibers. Fig. 6.13 shows an enlargement of the surface, which reveals its compli-

 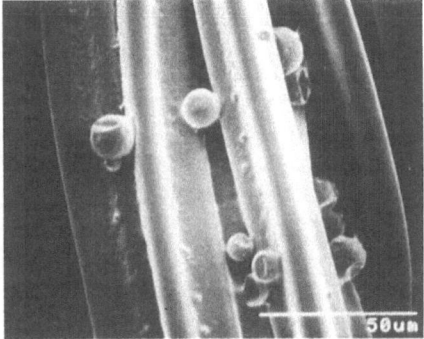

Fig. 6.14 Scented Textile (Source: Toray Industries, Inc.)

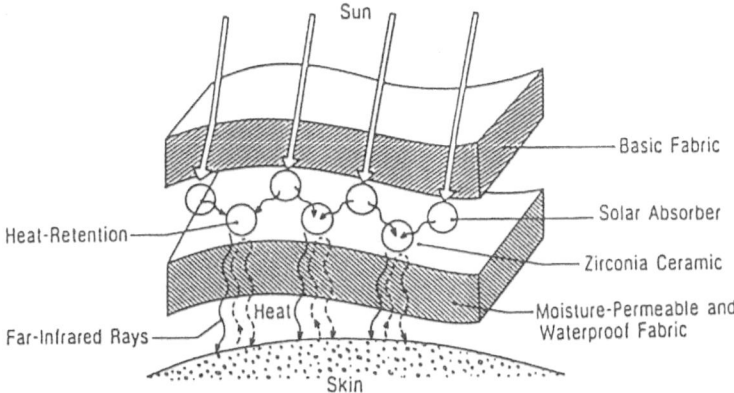

Fig. 6.15 Heat-retaining fabric

cated phases. It is a fabric consisting of multilayered heterogeneous fibers which are given a special processing treatment.

(5) Thermochromic fabric: "SWAY"

A recently highlighted interesting material is SWAY available from the Toray company. This is a temperature-sensitive product which changes color like a chameleon. In this product, fabrics are coated with thermochromic dyes contained in microcapsules. It is a product that not only stresses function but also appeals to consumer taste. The slide shows color changes under different temperatures. In this case, the pink one is at a temperature above 19 °C and the darker one is below 11 °C.

6) Scented textile

Scented textile contains microcapsules within the fibers diffuse jasmine, lemon and other fragrances as shown in Fig. 6.14.

(7) Heat-retaining textile

The principle of a heat-retaining textile is that, by absorbing sunlight, ceramics within the fibers emit far-infrared radiation to warm the wearer's body. Alternatively, absorbing body heat, they again emit far-infrared radiation as illustrated in Fig. 6.15.

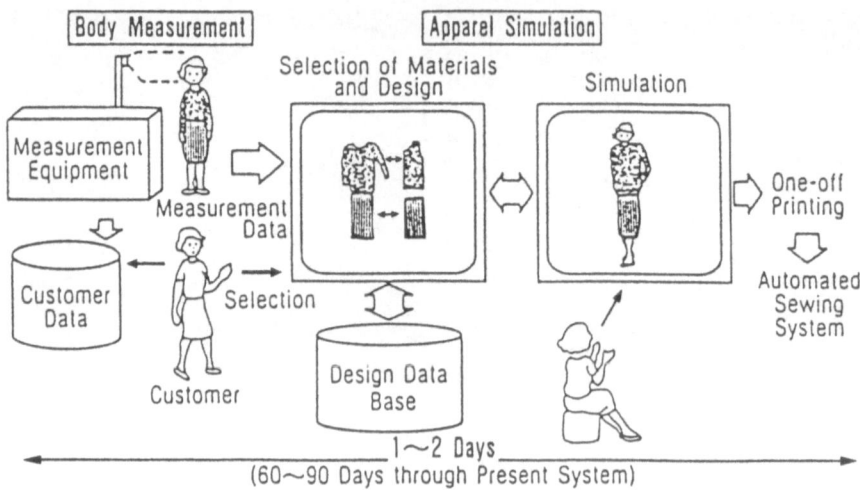

Fig. 6.16 Revolutionary system of apparel production

6.3.3
Innovation of Textile Production Process and Distribution Systems

6.3.3.1
Revolutionary System of Apparel Production

Three key words are important for new fiber technology, namely, high-technology, sensuousness, and quality. Crucial to the future development of the textile industry is the development of a supply system enabling prompt response to changing demand by better systematization aided by computer at the processing/distribution stages. As an example, Fig. 6.16 suggests upgrading apparel production processes. At present the apparel produstion process, from the selection of materials, color, patterns, etc. to manufacturing of products in various sizes, generally requires 60–90 days. This bold upgrade plan is designed to produce a garment which can satisfy a customer's requirements within a day or two, by three-dimensional size measuring, a dialogue with a data base, styling simulation, and dyeing, which will lead to a fully automated sewing system.

6.3.3.2
Fully Automated Sewing System

Because the development of this system not only requires huge funds but also takes a long time to develop, an industry-wide joint study is conducted as a national project in Japan. The project aims at developing technology to allow production of a wide variety of garments, each in small quantities, in less than half the time currently required. To this end, R&D efforts are being made to develop various basic sewing technologies as will as a total system to combine them all.

6.3.4
New Trends in Sensibility and New Synthetic Fibers

Polymer materials have recently been re-evaluated in Japan in – terms of sensibility, appeal to human interest in good sense – of apparel and high value added fabrics of so called "Shingosen" (new synthetic fibers). Out of the total of 15 billion dollars worth of annual textile imports, approximately 7 billion dollars worth are imported from Italy and France. Most of those textiles are of a good sense, and thus referred to as sensibility (Kansei) products, or sensible goods. Sensibility can be defined as follows.

1) Emotional capacity to respond to sensory stimuli.
2) The biggest part of human nature, yet unrefined even today.
3) An expression to summarize the incomprehensibility of consumer taste.
4) A key concept to understand a future trend of commercial products.
5) An antonym to RISEI (reason, i. e., an intellectual power to evaluate physical factors).
6) A sense common to the sense of the times and life.
7) Sensitivity to the significance of information.
8) An attractive condition in contrast to a necessary condition.
9) An aesthetic sense to external appearance and comfortableness (in the case of clothes).
10) Characteristics representing a symbolic value rather than a functional value.
11) A natural aptitude or good taste for finding an aesthetic value.
12) Mental and physical agreeableness.
13) A word which is employed to distinguish one's perspective view.
14) A capability to perceive the emotional change caused interactively.
15) Emotional wear in the sense that reason is a logical wear.
16) A movement of human mind (the sense of memory).

It may be critical in the following discussion to define explicitly what sensibility does imply. Sensibility is inherent to a human being, charaterized by its diversity resulted from environments, cultures, religions, laws, customs, etc., and thus concerned with a symbolic value of individuals. However, it may include many other functions behind our consciousness [6.8].

The author evaluates textiles in terms not only of the physical factors (reason) such as tensile strength and durability, but also of the mental factors (sensibility) – whether we like or not without reason. More recently, there appeared other factors to lead the fashion, which cannot be classified in two categories mentioned above. Those factors, which include exclusiveness, specialization, distinctiveness, sophistication, etc., represent the change of mentality by humans. Those factors can be summarized by the term wisdom, which may be defined as the intellectual capacity to create a unified concept upon the sensory experience received through sensibility as described in Fig. 6.17. Wisdom thus binds sensibility with intelligence, and acts as a communicator between a consumer who buys a product and a particular man or group who designs or shares it. Wisdom is able to create a new sense of value, so that a product should attract wisdom as well as sensibility and RISEI (reason) to be a good selling prospect [6-9].

Fig. 6.17 Schematic diagram
of tripolar relationship of reason, sensibility and wisdom
(knowledge)

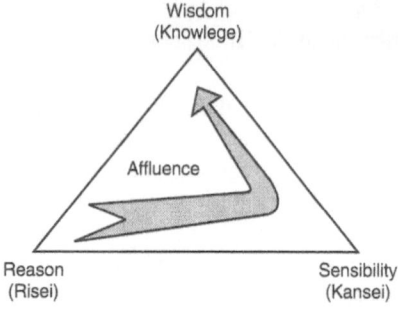

Human life is surrounded by polymers in clothes, interiors, paint, etc. The materials for fashion will be classified into two types; the materials created by mimicking the shape, structure and function of biomaterials in the stream of the biomimetics, and the materials created, not by mimicking but by surpassing.

The Shingosen (ultrafine polyester fiber)project was first started by mimicking silk, but went further to surpass silk by innovating an extremely fine filament, resulting in a completely new material which opened the possibility of creating a new generation of expression in the fashion world. Airplanes were first constructed by mimicking birds, but no one doubts that airplanes surpassed birds. As shown by these two examples, a real achievement is obtained when a mimic surpasses its original.

6.3.5
New Concept of Shingosen

Today, flexibility, imagination, comfortableness and quick response are four major human factors important in technology innovation, suggesting the time for brain power intelligence and wisdom has come. Reason is a pre-requisite for the development of polymer materials. Moreover, human factors such as reason, sensibility and wisdom play a vital role as a tripolar relationship in modern technology to create advanced industrial products as indicated in Fig. 6.14. If we are able to give even a slightly rational explanation to sensibility or wisdom, i. e., understanding of human brain power, that part of sensibility or knowledge can be incorporated into technology to produce value added specialty products. Sensibility and wisdom are tough subjects to analyze by the scientific approach of natural science.

People are forced to modify their view as natural science from a human viewpoint. For example, fuzzy and neuro are sophisticated terms often found in the new concept of design for new commercial products. Both terms, fuzzy and neuro belong to the arena of sensibility, although an explicit definition is not available. Although it is inter-disciplinary science these words, fuzzy or neuro, attract consumers since the words fill in the gap between reason and sensibility of human needs.

The concept of fractals, chaos and $1/f$ fluctuation can be applied to polymer technology including fibers and textiles, clothing, interior desgn, plastic products etc. to open a new prospect. For example, the plastic surface with a fractal structure affords more natural feeling, and printed wooden or marble pattern is an application of fractals. The $1/f$ fluctuation theory has been applied to translate paintings into music. As suggested in these examples, an individual sensibility can be developed and transformed into a sophisticated design applied to fiber technology. Utilization of brain power for innovation is essential for future technology.

6.4
High Performance Fibers and Advanced Composite Materials

6.4.1
Key Technologies of Super Fibers

Without a strong technological base, we cannot hope to satisfy the three key requirements, – high-technology, sensuousness and quality. Fortunately, as shown in Table 6.13, there is a variety of basic technology acquired through synthetic fiber business [6.10].

As an example of the development of high-value-added products, the following new fiber technology is reviewed.

(1) Super-fibers: "KEVLAR"

The best example c a high-tech product is "super-fiber" which challenges the utmost limits of strength. Typical examples are KEVLAR, aramid fiber of Du Pont Company and TORAYCA, carbon fiber of Toray. They form a critically important sector of the synthetic fiber industry for the future. As shown in Tables 6.13 and 6.14, aramid fiber clearly outpaces conventional ones in terms of both strength and modulus.

(2) Carbon fiber: "TORAYCA"

Toray is the world's largest carbon fiber manufacturer, carbon fibers being available in two types, PAN and pitch. Toray's TORAYCA is of the PAN type and claims a one-third share of the world market. It is a powerful, high-tech-based new material resulting from basic research.

Because it is widely used in such products as sporting goods, you may be familiar with this product. Lately, carbon fiber has also been used in aircraft. Thanks to its light weight, it is becoming an indispensable material for ensuring higher speeds and better fuel economy.

Table 6.13 "Super-Fibers" – Strength upon Strength

Comparisons of Fibers				
Materials	Tensile Strength		Tensile Modulus	
	G Pa	g/d	G Pa	g/d
Nylon	0.96	9.5	4.0	40.0
Polyester	1.16	9.5	12.2	100.0
Aramid	2.80	22.0	128.0	1000.0
Carbon	7.10	42.0	800.0	4755.0

Table 6.14 Comparison of physical properties in various high-tenacity high modulus fibers

Classification	Products	Manufacturer	Density	Strength (g/d)	Elongation (%)	Modulus (g/d)	Melting point ℃	Characteristics	Needs for Improvement
para-type aramide	KEVLAR® 29	DuPont	1.44	22	4.0	480	decomp.	–	–
	KEVLAR® 49		1.45	22	2.4	1,000			
	TWARON®	Euka	1.44	22	3.3	630	decomp.	–	–
	TECHNORA (HM-50)	Teijin	1.39	25	4.4	570	Ca. 500 decompl (N_2)	chemical stability wet-heat resistance fatigue resistance	–
poly-arylate	EKONOL®	Sumitomo Chemical Nippon Exlan	1.40	31	2.9	1,080	370	non water absorbency higher knot strength than Avamicle	adhesibility
	VECTRAN®	Kuraray Celanese	1.39	22 ~ 25	2 ~ 5	600 ~ 700	300	chemical stability non water absorbency	adhesibility
poly-ethylene	SPECTRA® 100	Allied	0.97	35	2.7	2,000	147	knot strength abrasion resistance chemical stability weather resistance	heat resistance adhesibility
	TECH-MIRONE ®	Mitsui Petro Chem.	0.96	35	3	1,160	146		ty
	DYNEMA®	Toyobo DSM.	0.98	30 ~ 55	2 ~ 5	800 ~ 1,400	146		
polyacetal	TENACK® SD	Asahi Chem.	1.41	8 ~ 12	5 ~ 10	160 ~ 320	180	low coefficient of linear expansion non water absorbency good chemical resistance	adhesibility heat resistance
carbon fiber	TORACA® T-800H	Toray	1.80	35 (570 kg/mm^2)	1.9	1,840 (30 T/mm^2)	–	strength under compression heat resistance	–

Source: Toray Industries, Inc.

6.4.2
Prospects for Carbon Fibers and CF Reinforced Composites

Composite materials acquire new capabilities by combining two or more materials, none of which themselves have these capabilities. An almost infinite number of composites could theoretically be created by combining existing raw materials. Composites have many practical uses; examples are macroscopic combinations, old plastics, yarn of comingled polyester and cotton fibers, laminated steel, self-reinforcing polymers, vulcanites, and alloys. But it is the development of carbon fiber that has made composites a significant high-performance material. In this section, we discuss the future of advanced composite materials (ACM). Reinforced mainly by carbon fibers, ACM are considered to be a promising new material [6.11].

Twenty-two PAN-based and six pitch-based carbon fiber manufacturers now supply their products to the entire world. As Fig. 6.18 indicates, their most promising markets for the next five years will be in Europe and the United States. The success of these enterprises therefore depends to a large extent upon their penetrating European and American aerospace and aircraft projects.

Many of the relevant firms in Europe and the United States have recently been reorganized. The era when profits could be reaped merely by producing carbon fibers or prepregs has come to an end. The currently accepted concept holds that those companies will succeed that are able to invest huge sums in long term R&D, while integrating the process of transforming raw materials into final products.

Since there are many carbon fiber makers compared to demand from the aerospace industry, competition for participation in aerospace project is severe. In order to succeed under these conditions, well balanced all round technologies are essential; it should utilize matrix resins, design, molding technology, and reinforcing fibers. Only about half a dozen companies will survive to enjoy the benefits of the real age of composite materials.

Fig. 6.18 The change of world market in ACM (Advanced Composite Materials)

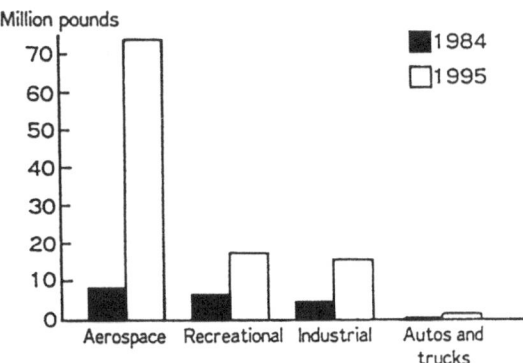

6.4.3
Trends in Technological Development of Carbon Fibers

Why is carbon fiber so strong? The reason lies in the hexagonal basal plane, which is formed by the carbon atoms that make up carbon fiber; it is parallel to the fiber axis. Properties of quasi-isotropic CFRP are illustrated in Table 6.17. In theory, the stress needed to break the carbon-carbon bond in the graphite layers is up to 200 GPa.

However, the strength of PAN-based carbon fiber now available is only 3–5 GPa. Experiments show that the tensile strength of carbon fiber increases when its gage length is decreased. If this finding is extrapolated, the tensile strength that could ultimately be attained is about 8–10 GPa; this is far below theoretical limits. Although it is difficult to approach these limits at present by attaining perfect molecular and crystalline strength, strength can be enhanced by removing such macroscopic defects as impurities, foreign objects, and voids that exist within and outside the fiber. Up to 7 GPa is now attainable through strict process control, such as complete cleaning of precursors, prevention of fiber melting in the carbonization process, and prevention of tar adhesion by removing off-gas. The development of improved physical properties is illustrated in Fig. 6.19.

In a broad sense, the quality of products is mainly determined by technolgy, but price is not solely determined by technology. Price is very difficult to forecast, because it is affected not only by cost reduction from new materials technology and by improved production processes, but also by value and by supply and demand. As for carbon fiber, it seems unlikely that new production techniques in which oxidation, carbonization, and graphitization differ basically from current techniques will be forthcoming in the near future. A more realistic possibility is that cheaper raw materials will be used.

Fig. 6.20 shows an example of the constituents of production costs. It reveals that precursor costs are a very high percentage of total costs. It follows that pitch-based carbon is likely to be developed. Raw pitch is undoubtedly much cheaper than PAN as

Fig. 6.19 Direction of carbon fiber improvement (Source: Toray Industries, Inc.)

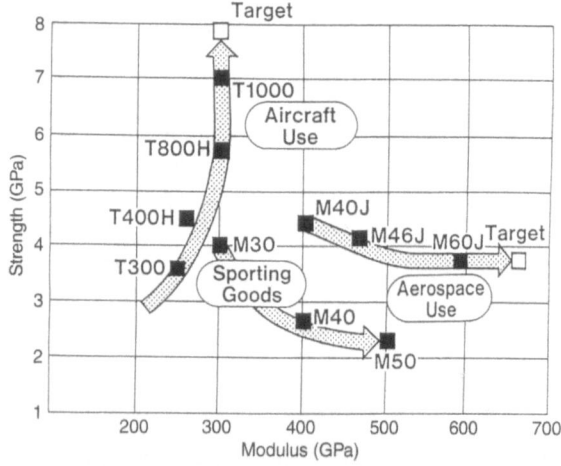

Fig. 6.20 The items of production cost of PAN carbon fiber (example in RK30)
A: precursor 36 %
B: marketing and quality management 17 %
C: depreciation 16 %
D: labor cost 11 %
E: chemical cost 11 %
F: carbonization and oxidation 9 %
(Source: Toray Industries, Inc.)

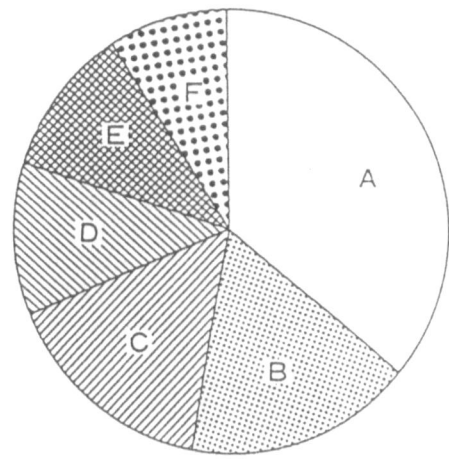

an initial component; it also has a high yield for carbonization. However, since pitch would have to be refined by heat treatment to make the performance of pitch-based carbon fiber comparable to that of fiber based on PAN, the yield would decrease. As the melt viscosity of pitch is highly dependent on temperature, care must be taken to spin pitch steadily. The precursor is brittle; it must therefore be handled carefully.

In due course, these problems will probably be solved, and a high-performance precursor from pitch is likely to become cheaper than that based on PAN. On the other hand, short pitch-based fiber with moderate properties for general use will probably be developed. Indeed, this type of carbon fiber is already being produced by Amoco Performance Products, Inc. through Toray CF technology and Kureha. It is widely used for many purposes, mainly asbestos-substitution and concrete-reinforcement. It follows that prospects are excellent for industries that enjoy easy access to supplies of pitch resins ...

6.4.4
Matrix Resins

From the time that glass fiber was predominantly used for reinforcement, unsaturated polyester and vinylester resin systems have been widely utilized as matrix resins. Advanced composite materials (ACM), especially carbon fiber reinforced composites (CFRP), are currently dominated by epoxy matrix resins, which show excellent heat resistance, mechanical properties, and environmental durability. In order to extend their use in aircraft, both fiber-oriented and non-fiber directional properties should be improved. In addition, tensile strength, compressive strength, damage tolerance (residual compressive strength after impact), and hot/wet properties are important, but they cannot simply be achieved by improving the properties of fiber.

Technology to toughen the matrix resin system is badly needed, and it has been the subject of research in many parts of the world. At the first stage of study, the chemical structure of epoxy resin was specially designed to improve its elongation at failure.

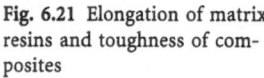

Fig. 6.21 Elongation of matrix resins and toughness of composites

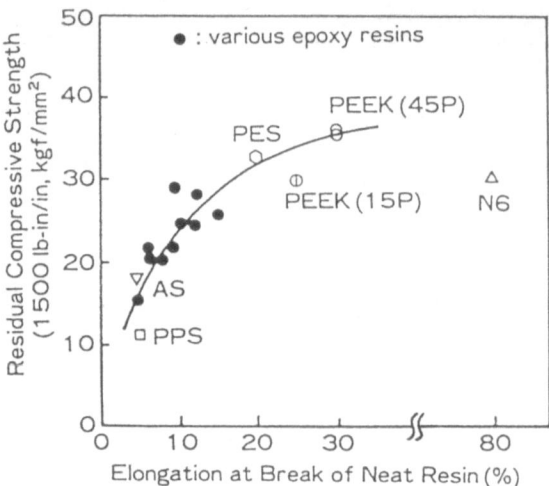

But this approach has reduced the glass transition temperature of the systems. Recent work in this area has concentrated largely on toughening epoxy materials by incorporating thermoplastics.

On the other hand, heat resistance is achieved by thermosetting. In this case, it is important to achieve optimum morphology by controlling the compatability of both components.

Elongation of matrix resins and toughness of composites using new polymers such as PPS(polyphenylene-sulfide), PEEK(polyether-ether-ketone), PES(polyester-sulpnone), and a variety of epoxy resins are illustrated in Fig. 6.21.

It has been reported that tough resin systems have optimum two-phase microstructure with the modifier present in discrete domains; this is achieved by controlling the compatibility of epoxy/thermoplastic resins. Epoxy resin does not have the heat resistance needed in more severe conditions, e.g., as structural materials for aerospace or materials for engines.

In response to demand for high temperature polymers, thermosetting polyimide resins have been developed by NASA.

But these resins do not have sufficient processability, due to their high cure temperature, high viscosity, and the existence of residual solvents. Bismaleimide resins and oligomeric imide have been developed, but they are not yet sufficiently processible. A study of polyquinoxaline resins has been completed by the Research and Development Institute for Metals and Composites for Future Industries, sponsored by the MITI Agency for Industrial Science and Technology (Table 6.14). The representative polyimide resins for matrix applications are summarized in Table 6.15.

On the other hand, thermoplastic resins have many advantages as matrices as illustrated in Table 6.16. Their use seems to be rapidly becoming more widespread. As compared with thermosetting resins, thermosplastic resins achieve excellent impact resistance. The elongation of matrix resins and toughness of composites are described in Fig. 6.21.

Table 6.15 Representative polyimide resins

resin name	monomer structure			cure temperature (°C)	aftercure temperature (°C)	T_g (°C)
	end-structure	tetracarboxylic acid	diamine			
PMR-15[18]				316	343	345
LARC-160[19]				316	316	325
Theroid 500[20]				260	271	340

Table 6.16 Properties of thermoplastic matrix composites

advantages	disadvantages
high toughness	difficult impregnation
high cycle processability	high temperature-high pressure process
hot-melt adhesion, post-forming, recycle easy	low heat resistance: low Tg, low Tm
storage of prepreg	non-tacky

6.4.5
Properties of Composite Materials

The most outstanding advantage of carbon fiber reinforced composites is their strength and high specific modulus. However, this is not to say that these properties are fully exploited. The fundamental design concept for composites is to take advantage of their anisotropic properties, but, in many practical applications, a quasiisotropic laminate (form 0/45/90) is used instead. Doing so causes some problems. Thus, since composites do not yield like metal, the strength of the laminate decreases dramatically due to the open-hole. Furthermore, the compressive strength of composites decrease greatly, because of the transverse crack or delamination due to the low fracture toughness of the interlaminar layer and the transverse direction of the fiber.

Composite materials permit the designer much freedom, but using this freedom is not easy. The freedom is of recent vintage; it results from these materials being 30 % lighter than aluminum alloys and from the development of high-strength fibers and toughened resin systems as shown in Table 6.17. However, the increase in compressive strength is much smaller than the increase in tensile strength. Thus, improving

Table 6.17 Properties of quasi-isotropic CFRP laminate [0/45/90]$_s$ (compared to aluminium alloy)

Materials	CFRP		Aluminium alloy (7075–T6)	Necessary value for 30 % lighter than aluminium alloy*
	T300/ Conventional resin	T800/ Toughened resin		
Specific gravity	1.6	1.6	2.8	1.6
Elastic malleability (Gpa)	52	62	70	57
Tensile strength (Mpa)	560	840	550	490
Open-hole tensile strength (Mpa)	330	670	470	380
Compressive strength (Mpa)	670	700	450	370
Open-hole compressive strength (Mpa)	360	380	400	330
Compressive strength after impact (Mpa)	160	350	450	370

$$* = \frac{1.6}{2.8} \times \frac{1}{1-0.3}$$

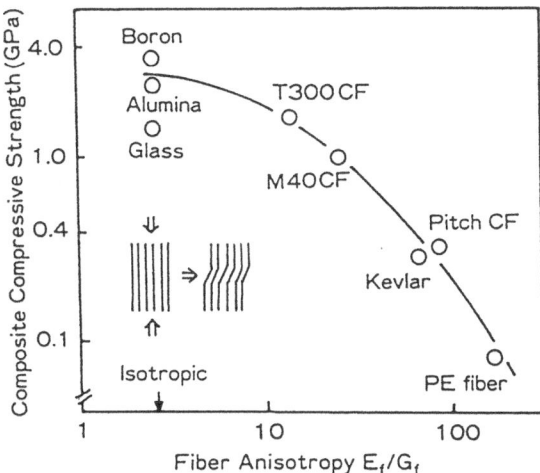

Fig. 6.22 Dependence of composite compressive strength on fiber anisotropy (Source: Toray Industries, Inc.)

compressive strength remains an important problem. Contrary to the tensile load, the compressive load cannot be supported by the fiber alone; for this reason, rigid matrix resins are needed to prevent buckling of the fibers and to enhance compressive strength. It follows that the development of a high modulus resin system, even under hot/wet conditions, is essential to finding a solution of this problem.

The compressive strength of fiber reinforced composites seems to be related to the anisotropic parameter of the reinforcing fibers E_f/G_f. The tensile modulus and shear modulus are shown in Fig. 6.22. It is necessary to develop a carbon fiber that has high compressive strength by optimizing fiber microstructures.

6.4.6
Molding Technologies

Molding technologies for ACMs have not been improved as much as those for glass fibers. To extend the areas where carbon fibers are used, sophisticated high-productivity molding systems should be developed by exploiting the capabilities of computers and robots. Equally important would be such technological breakthroughs as developing a molding system for thermoplastic composites.

If ACMs are to be used by the automobile industry, where demand is enormous, new design concepts will be necessary, such as integrating many parts into a single component. To fulfill these requirements, a molding technology should be developed that can produce complicated and integrated structural parts. The basic technology will be conventional resin injection and high-speed stamping, modified accordingly.

FRM (Fiber Reinforced Metal) is the series of composites in which the matrix metals are reinforced by advanced fibers. Carbon fiber is studied as one of the reinforcing fibers. The carbon fiber/aluminum system is of particular interest.

The difficulties are poor wettability and the reaction between fiber and matrix. However with pretreatments of carbon fiber were studied, each method has both merits and demerits, and no solution has been established yet. In continuous fiber reinforced FRM, the fiber strength can be utilized efficiently. However, with short fiber such as alumina fiber and silicon carbide whisker reinforced FRM, although it can be fabricated easily by casting methods, the reinforcing effect is not sufficient. Since the market of FRM is mainly for space applications, it is not yet very advanced in Japan. This material is also involved in the above mentioned R&D project sponsored by MITI [6.10].

Since carbon fiber has a number of unique properties, besides its general properties such as light weight and high strength, properties such as electric conductivity and X-ray transparency have been applied to antennas and brushes to remove electric charges, electromagnetic shielding and X-ray cassettes. Speaker diaphragms with high modulus carbon fiber combined with a highly damping resin system, are widely used, the application being the third biggest market following sporting goods and aircraft structural applications.

One of the major uses of carbon materials is in the electrodes of batteries having good chemical resistance and electrical conductivity. Electrodes of sheet form are necessary for the new battery system in the Moonlight Project promoted by the Japanese NEDO (New Energy Development Organization). The solar cell is used in various applications, from calculator to artificial satellite. Since conversion efficiency of amorphous silicon is limited, increasing the cell area is one solution to increase capacity. Carbon fiber is the candidate for this application. Nuclear fusion is another application as the artificial sun needs a superconductive magnet to shut in the plasma.

The method to vapor-deposit niobium carbon nitride on to carbon fibers has been studied in Germany to make a superconductor which has high critical temperature, high critical magnetic field and high endurance against neutron radiation. Carbon fiber has anisotropy in thermal conductivity, based on the fact that the graphite plane is oriented parallel to the fiber axis. Carbon fiber has so-called semiconductivity, in which electrical conductivity changes depending on temperature and humidity.

6.4.7
Lesson of Fiberglass Reinforced Composites

Glass fiber was developed in the 1930s. During World War II, it was used for such military purposes as aircraft radar. After the war, when glass fiber was used for fishing rods, the industry expanded rapidly. However, it took 50 years before plexiglass was used for the Corvette sports car. Only 20 years have passed since carbon fiber was developed. Europe and the United States use it for military purposes, and Japan does so for fishing rods and golf clubs. If history repeats itself, large quantities of carbon fiber will be used by the automobile industry in the twenty-first century. Although the mass media and academia have been aware of carbon fibers for a long time, many problems remain unsolved, e. g., improving the properties of carbon fiber and matrix resin, utilization of anisotropy, reliability of materials, and high speed molding technologies.

When these problems have been overcome, society will acknowledge the importance of ACMs. Huge chemical companies are starting to make large investments in this

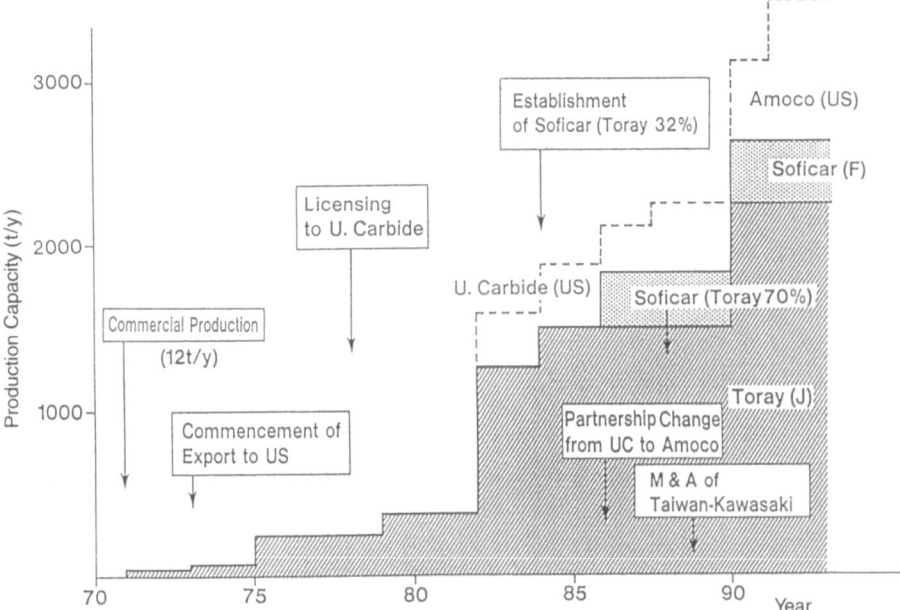

Fig. 6.23 Expansion of carbon fiber business in Toray
(Source: Toray Industries, Inc.)

field. Since Japanese industrial technology is still vulnerable, we will be left behind in this area unless government, industry, and academia work together on the R&D of composites. Success will depend on performance improvement, proprietary position, personnel development, price reduction, and, above all, patience. The expansion of Toray's carbon fiber business from the start of commercial production in 1971 to 1992 is shown in Fig. 6.23.

6.5
Plastic Optical Fibers for the Communication Industry

6.5.1
Overview of Fiber Optics

Plastic optical fibers (OF) are already in wide use in a large number of fields and both the range of applications and the market are expected to continue to expand steadily. Properties of plastics OF and silica OF are compared in Table 6.18. Transmission loss of plastics OF is 500times larger than that of silica OF, however, some merit of plastic fiber is recognized such as light weight and mechanical flexibility. The trends of a variety of plastic optical fibers and main applications with Japanese manufacturers are shown in Table 6. 19.

The progress and current technology trends of plastic optical fibers in Japan are reviewed in terms of structure and materials for plastic optical fibers (POF). Progress

Table 6.18 Plastic OF vs. silica OF

	POF	Silica OF
Trans. loss (dB/ka)	125 (650 nm)	0.2 ~ (1300 nm)
Bandwidth	~ 10 MHz · ka	10000 MH · ka
N. A	0.3 ~ 0.6	0.1 ~ 0.25
Available wavelength	Visible	Visible ~ Infrared
Mechanical properties	Better Flexibility	Brittle
Fiber diameter (Bulk Fiber)	100 ~ 3000 μ	100 ~ 500 μ
Processability	Easier handling	Special tooling
Heat resist (dry)	< 85 °C	150 °C
Chemical resist.	poor	good
Specific gravity	1.2 ~	2.4 ~
Cost (Bulk, System)	Cheap	Expensive

Table 6.19 Types of plastic optical fibers

Core		Cladding	Main Application	Maker (in Japan)
Thermoplastic	PMMA	Fluoropolymer	Data-trans., Sensor Lightguide Sign	Mitsubishi Rayon Asahi Chemical Toray
	PSt	PMMA	Display	
	PC	F-polymer (Polyolefine)	Heat resist. Data-trans., Sensor	Mitsubishi Rayon Fujitsu, Teijin Idemitsu, Asahi
Thermosetting	Polysiloxane	F-polymer	Heat resist. sensor	Sumitomo Electric
	Crosslinked structure	F-polymer	Heat resist. sensor	Hitachi

has been made in lowering the transmission loss of optical fibers. The improvement of heat resistant plastic optical fibers and development of the multi fiber endoscope and several other items are observed. The scope of the development of plastic optical fiber technology in Japan has been reviewed [6.12].

6.5.2
Structure of Plastic Optical Fibers and Distinctive Features of New Products

Plastic optical fibers (POF) have a plastic core and cladding, and all the commercially available ones are of the multi-mode step-index type. The graded-index type is not on the market yet. The diameter of POF is roughly 3-10 times larger than that of silica ones, but, even though they are larger, they are much more flexible and much tougher, which is one of the advantages of plastic. The cladding layer is far thinner than the

Table 6.20 Properties of optical plastics

	Refractive Index N_D	Abbe No. V_D	Trans- mittance (χ)	Working temp. (°C)	Specific gravity (g/cc)	Thermal expansion (m/m °C) x 10^{-6}
PSt	1.591	30.9	90	70	1.06	80
PC	1.586	30.3	89	120	1.20	70
CR – 39	1.498	57.8	92	70	1.32	117
PMMA	1.492	57.2	92	70 ～ 100	1.19	63
TPX*	1.466	56.4	–	180	0.87	117

*: poly-4 methl-pentene-1

core diameter. This is another feature which distinguishes plastic from silica fibers. POF, which can be used to transmit optical signals along plastic fibers, are in many ways superior to their glass counterpart. Table 6.19 shows the comparison of the properties of POF with silica optical fibers. They are easier to handle with excellent ductility and light weight, easier to splice together and to light sources because of their large core diameter and high numerical aperture (NA).

The characteristic feature which distinguishes POF from silica fiber is the fact that amorphous polymers sustain much higher strains before suffering permanent damage in contrast to the brittle behavior of silica. These properties permit fibers of large diameter to be bent to tight radii of curvature without damage. The properties of optical plastics are summarized in Table 6.20. Presently available fibers constructed from PMMA are available at around 1 mm diameter which can sustain bend radii of less than 10 mm.

6.5.3
Lowering Transmission Loss

The most important challenge to POF for application to long distance communications is how to reduce its attenuation. The improved version of PMMA core shows transmission losses of less than 125 dB/Km at 650 nm wavelength.

Considering the developing stage of industrial manufacturing, the attenuation obtained has reached to a reasonable level compared with the loss limit of 106 dB/ Km.

Progress has been made in lowering the transmission loss of PMMA core type of fiber well balanced between performance and cost. Transmission loss is one of the most important measures of the performance of optical fibers.

In Table 6.21, structural factors causing transmission loss are shown in terms of inherent and external factors. The inherent absorption and scattering arise basically from the molecular structure of the core polymer. They are much more serious in plastic fiber than in silica fiber. In material with C-H bonds, the higher harmonics of the vibration of this bond have effects that are felt all the way up to the visible region. Transmission loss of PMMA core POF is composed of inherent absorption due to carbon hydrogen (C-H) vibrations and the electronic transition, the intrinsic Rayleigh

Table 6.21 Factors causing transmission loss

Inherent factors	Absorption	• Higher harmonics of C-H vibration ◦ Electronic transition
	Scattering	◦ Rayleigh scattering
External factors	Absorption	◦ Transition metals ◦ Organic contaminants ◦ OH group
	Scattering	• Dust and Micro void ◦ Disturbance at core/clad. interface ◦ Fluctuation of core diameter ◦ Microbending ◦ Birefringence due to orientation

scattering due to imperfections in the waveguide structure such as core diameter fluctuations as well as the particulates and other impurities from the environment.

6.6
Management of the Fiber Industry in the Age of High Technology

6.6.1
Worldwide Operations

The author has considered the modernization and remarkable progress of Japanese synthetic fiber technology and high value-added Shingosen industry since Meiji restoration in 1886. When the author looks into the trends of Japanese fiber industry in a long tradition, the author realizes that the modern Japanese fiber technology of Shingosen is actually based on the sensibility of Japanese people. It has been nurtured by the old traditional core fiber technology accumulated through production know-how and processing wisdom of natural fibers such as natural silk and cotton as well as natural weaving and dyeing with natural dyes. The authentic textile production technology is already found in the Nara period of the sixth century in Shosoin Temple treasury exhibits (Gyobutsu).

As to the global fiber business in the international market, the only way that the Japanese fiber and textile industry can survive in the twenty-first century is to increase productive performance by optimizing the integrated total cost of production in terms of social cost of infrastructure such as pre-production cost, production cost at the plant site, and post-production cost such as recycling product liability and after-service including social cost. Tc, be competitive in the international market, the development of new value-added products together with both sophisticated material and design in fibers, textile and apparel are indispensable to solve the issue of hollowing out of fiber industry in domestic market. Basic research and process development should be emphasized in interdisciplinary collaboration of material suppliers, fabricators, converters and apparel designers. Process automation by introducing computer technology is essential to enhance productivity of the fiber and apparel industries. New distribution systems and quick response of variety of products to customer needs should be established. Speaking of Japan, the unique and influential international distribution mechanism based on customary transactions in the domestic market does not apply to a borderless system of economy. This is the weakness of Japan. In 1989, MIT reported the comparative study of the United States industry by the Commission of Industrial Productivity (JCIP) with 60 academic professionals. They conducted 500 – 600 interviews and visited 200 companies in 8 industrial fields for two and a half years and presented more than 10 proposals to bring these strengths and weaknesses more into balance. Responding to the challenge, the author has taken initiative in Japan to set up "the Japanese Commission of Industrial Performance" in 1990. [6.13] After three years of extensive studies in seven different Japanese manufacturing business sectors, together with 16 professors in academia and 34 business corporations, we published the report entitled, "Made in Japan," by Diamond Publishers, Tokyo in 1994 [6.14].

Manufacturing is a major industry in both countries and its value-added ratio is 28.2 % in the United States and 34.2 % in Japan. The author agrees with the common saying that "Prosperity of a nation depends on its excellent manufacturing capability. What the author wants to emphasize here is that Japan should recognize the fact that international communities are waiting to see whether Japan can change herself in the paradigm shift into a nation of contributing to the world with a "give and take" attitude.

As mentioned earlier, Japan is now facing a turning point of paradigm shift from confrontation to cooperation in the new concept of *economic symbiosis*, that is interdependence with each other and global coexistence and role-sharing in the global environment. Transfer of technology to developing countries in NIES and ASEAN economic regions and Pan-Pacific rim countries including mainland China will become increasingly important.

The Japanese synthetic fiber industry must take greater roles and responsibilities through research and development of new synthetic fibers and other high technologies, improvement of productivity by technological innovation, and technology transfer. Soviet and many East European countries have begun to work with the same language and doctrine, beyond conventional allied relations. If Japan continues to act against international sympathies, it will be difficult to restructure reliable relationships.

Japanese industries must not be isolated by going against the movement of globalism. We must immediately set our own vision for contributing to the world, review our subjects, and promote mutual understanding among the United States, Japan, and Europe. Now is the time for Japan to show her willingness to work for the world.

6.6.2
Management in High-Tech Super-Industrial Society

From the viewpoint of corporate management, internationalization or globalization is today a matter of increasing importance. The conventional pattern, where products are manufactured in Japan alone and then exported, is no longer possible. The author

Table 2.22 Management in the Age of High Technology

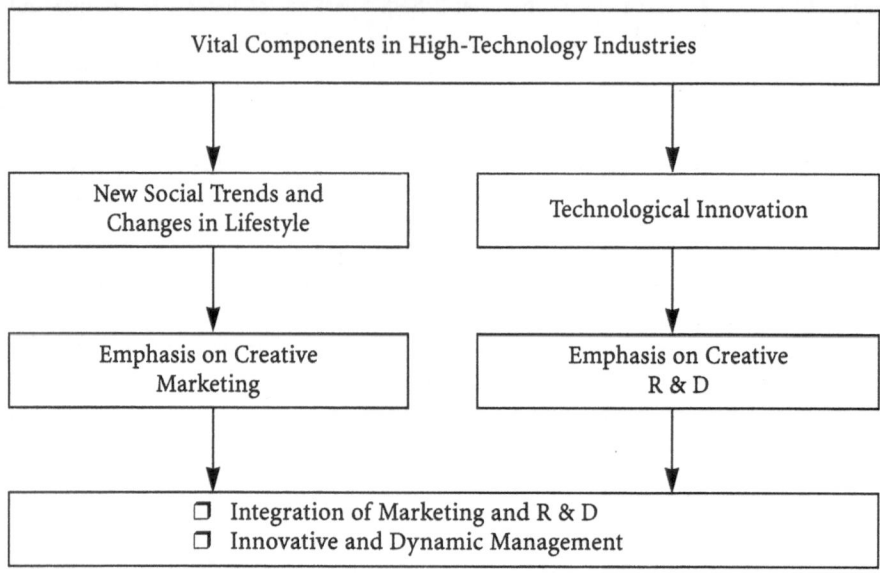

thinks that the so-called borderless economy is becoming a reality. In such an age, the secret of business expansion is that each player correctly recognizes and fulfills the roles it is expected to play. Based on such a recognition, Toray has long been committed to international operations.

Finally, the author would like to comment on the desirable management system in fiber industry. It is important to emphasize both marketing and R&D in an effort to cope with a fundamentally changing social structure of fiber business as shown in Table 6.22. At the same time, it is vital to stress strategic marketing. In order to avoid hollowing out of the Japanese industry, it is believed that the prerequisites for high-tech industry are flexibility to cope with the new social trends and a new lifestyle and technological innovation. These require executives to make a commitment to creativity. The ultimate goal is to become a "winner in the market" with the competitive advantage.

In drastically changing markets, "marketing conceived by consumers" alone is no longer sufficient. As a matter of course, such a concept must be backed by the creation of new products to help steer the market.

The author's cherished concept is the importance of social contract and that the strategic marketing will enter a new age only when consumer-based marketing and R&D efforts are united hand-in-hand. Taking the example of a vehicle, the front wheels of "marketing" and the rear wheels of "R&D" must work together well as a four-wheel mechanism to achieve symbiosis in international community. The concept is shown in a diagram of Table 6.23. Manufacturing is a major industry in both countries and its value-added ratio is 28.2 % in the United States and 34.2 % in Japan.

Table 2.23 Stress on Strategic Marketing

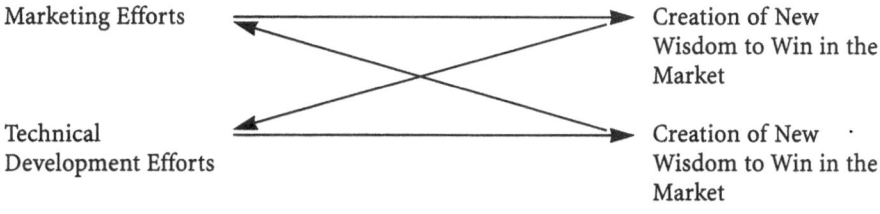

Goal of the Marketing ······ "To Win in the Market"

Marketing Efforts	Creation of New Wisdom to Win in the Market
Technical Development Efforts	Creation of New Wisdom to Win in the Market

Marketing and R & are the Front and Rear Wheels of the Car

"Prosperity of a nation depends on its excellent manufacturing power. What the author want to emphasize is that Japan should recognize the fact that world countries are waiting to see whether Japan can change itself into a nation working for the world.

Japan is now facing a dilemma of harmonizing contradictions of *competition and cooperation* to materialize international coexistence and role-sharing. To produce a vision of globalizing the Japanese economy, Japan has recently set up the Japanese Commision on Industrial Performance."(JCIP). Within the organization, the JCIP Survey Committee has started its activities [6.1], and the report was published in 1994 including study of fiber and textile industries [6.14].

In this respect, the new concept of C&C, i. e., competition and cooperation, is important. It is necessary to establish the new system of international cooperation such as long range, basic R&D among industry, academia and national laboratories in precompetitive stages.

The author sincerely hopes that the future survey activities will be successful to revitalize Japanese industry, so that it will become truly respected by the international community. Furthermore, the author is fully convinced that the comparative study of industrial performance in advanced societies including the fiber industry, will reveal insight into the management strategy for mutual coexistence and *economic symbiosis* in the global market.

Acknowledgement. The author is indebted to Mr. A. Higuchi, Director of Fibers and Textiles Research Laboratories, Toray Industries, Inc. for technical informations on the synthetic fiber industry.

6.7
References

6.1   Yoda N (1991) Japan Overcomes Natural Resources.
         Do Not Be Arrogant, (Japanese) Keizaikai, Tokyo (English version forthcoming)
         Yoda N (1989) TBR (Toray Corporate Business Research, Inc.) Intelligence, 1:2 48; Yoda N(1990)

TBR Intelligence 2:2 58; Yoda N (1990) TBR Intelligence 3:1 23; Yoda N (1991) TBR Intelligence 3:1 38; Yoda N (1991) TBR Intelligence: 3:2 16; Yoda N (1992) TBR Intelligence 4:1 15.

6.2   Svedova J (ed) (1990) Industrial Textiles Text Sci and Tech 9:15.

6.3   Verret R (1991) J Text Inst 82: 129–131.

6.4   Taniguchi F (1991) J Text Inst 82: 195.

6.5   Matsumoto A (1987) Japan Fibers Assoc Preprints of New Fibers 2.

6.6   Hayakawa A (1986) J Text Fibers Assoc 42: 328.

6.7   Kobayashi S (1991) in: Preprints International Chemical Fibers Congress Dornbrin Austria.

6.8   Okamura S (1991) in: Preprints 2nd Pacific Polymer Conference: 343 Otsu Japan.

6.9   Okamoto M (1991) in: Preprints 2nd Pacific Polymer Conference: 349 Otsu Japan.

6.10 Yoda N (1989) TBR Intelligence 1:2 48.

6.11 Aotani H (1989) TBR Intelligence 2:1 26.

6.12 Ide F, Hasegawa A (1991) in: Preprints of Indian Emerging Frontier Polymer Conference New Delhi, India.

6.13 Yoda N (1990) MIT Report Made in America (Japanese translation) Soshisha Tokyo.

6.14 Yoda N (1994) in: Yoshikawa H (ed) Made in Japan Diamond Tokyo.

Chapter 7

Recent Progress in the Objective Measurement of Fabric Hand

Sueo Kawabata and Masako Niwa

7.1
Introduction

There are many materials which are used close to humans and cause interactions between the material property and human sensibility. Polymers are such materials, a typical example being clothing material made of organic fibers (Figure 7.1). Although the strength and durability of the material are important in utility performance, the fitting of material properties with humans is more important with clothing material.

Consider two table plates, one made of wood and another of steel or plastic. Most people would prefer the wooden plate for everyday use because of the better feel of the wood surface. A material which is used in circumstances near humans and evaluated by humans as fit for them is called "human material" by the present authors.

In these human materials, fitness for humans is a primary performance as well as utility performances such as strength, durability etc. However the evaluation method for material fitness for humans has not been systematically investigated and has been carried out only by subjective methods by means of human hands and fingers; this has been called "hand evaluation".

In this chapter, the analysis of the subjective hand evaluation of clothing material is first introduced, and then a trial aimed at the objective evaluation of the fabric hand is introduced. These researches have been developed over the past 20 years by the present authors and their colleagues.

Fig. 7.1 The human materials

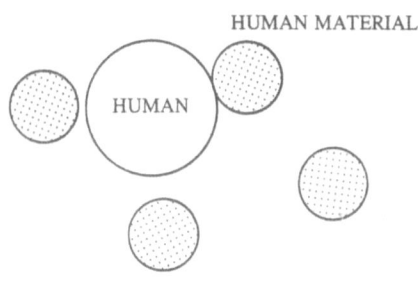

Okamura, Rånby, Ito (Eds.): Macromolecular Concept and Strategy
© Springer-Verlag Berlin · Heidelberg 1996

7.2
Hand Judgement for the Evaluation of Fabric Character and Quality

There are two types of performances for clothing fabrics [7.1]. One of them is utility performance such as strength, durability, dimensional shrinkage caused by washing etc. These factors are important for clothing material. After this requirement has been satisfied to some extent, consumers seek better quality from the comfortable wearing point of view. This performance in relation to "better-fitting" to the human body is the next, essential requirement of clothing material. The evaluation of the fabric quality which is involved in this second performance is, however, not as simple as the evaluation of the utility performance.

The quality of clothing fabric concerning the second requirement has been evaluated by consumers and textile producers subjectively by means of hand touch of fabric from the mechanical comfort viewpoint. This evaluation has been called "hand evaluation" and the fabric property relating to this evaluation "fabric hand" (handle in England). The subjective judgement of fabric hand is based on human sensitivity and experience. It is true that this subjective method is the most direct method to evaluate the fabric mechanical comfort, because the sensitivity of the human body feels the comfort of clothing.

Although this subjective evaluation is essential and highly sensitive with the accumulation of experience, there is a problem in its subjectivity which restricts the scientific understanding of fabric hand for those who intend to design high quality fabrics by means of engineering.

Because of the importance of the scientific understanding of fabric hand, many trials for replacing the subjective method with an objective method have been carried out by many researchers in the textile field since the trials by Peirce [7.2] in 1930, which suggested the correlation between fabric hand and fabric mechanical properties. The trend of research in this field, however, was directed to the explanation of fabric mechanical properties rather than the fabric hand itself. However the importance of the understanding of fabric hand has been considered throughout.

The present authors initiated research on fabric hand around 1968 in Japan on the basis of their concepts of fabric hand and the basic research on fabric mechanical

Fig. 7.2 Fabric hand evaluation of high quality men's suiting

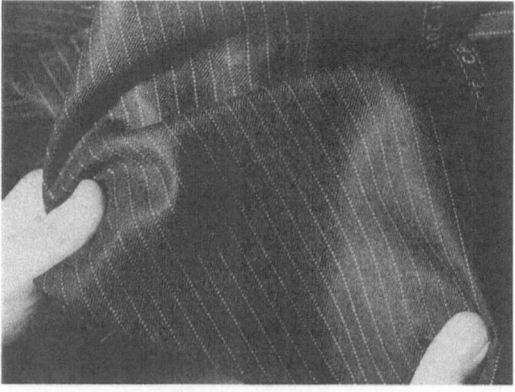

properties which had been carried out by them [7.3]. The target of the research was focussed on the analysis of the judgement of fabric hand as carried out by experts in textile mills, especially finishing mills for wool textiles. The first step of the research was to standardize the fabric hand expression which had been used by the experts in wool textile mills. The Hand Evaluation and Standardization Committee (HESC) was organized by Kawabata, one of the present authors, in 1972 and about twelve of the experts from the mills joined the committee at that time.

7.3
Primary Hand and Total Hand

During preliminary discussions with the experts, the way they judged fabric hand was surveyed. Figure 7.2 shows the hand judgement of a fabric by an expert who is getting a lot of information about the fabric character as concerns the fitting to humans. He then judges the quality of fabric as to its good or poor hand.

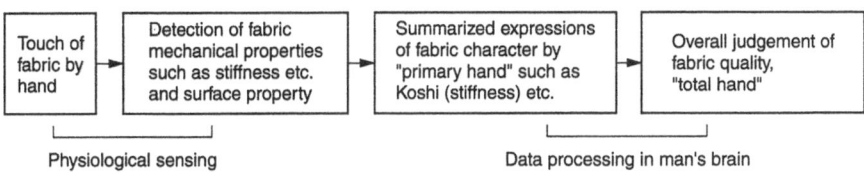

Fig. 7.3 The sequence of the subjective judgement of fabric hand by experts

Table 7.1. The primary hands

Men's suiting	
Winter suiting	Summer suiting
1. Stiffness (KOSHI)	1. Stiffness (KOSHI)
2. Smoothness (NUMERI)	2. Crispness (SHARI)
3. Fullness (FUKURAMI)	3. Anti-drape (HARI)
	4. Fullness (FUKURAMI)

Women's thin-dress fabrics

1. Stiffness (KOSHI)
2. Anti-drape stiffness (HARI)
3. Crispness (SHARI)
4. Fullness (FUKURAMI)
5. Scrooping feeling (KISHIMI)
6. Flexibility with soft feeling (SINAYAKASA)[a]
7. Soft feeling (SOFUTOSA)[a]
8. Smoothness (NUMERI)[b]

[a] The SHINAYAKASA and SOFUTOSA are not primary hand but semi-total hand, these were added here as primary hand because of their frequent use in expressing fabric character
[b] Smoothness (NUMERI) is also used in some cases

Figure 7.3 shows the sequence of the judgement of the experts [7.1]. The first step is to touch a fabric by hand and to detect the mechanical properties perceived by human sensitivity.

The first judgement is done to evaluate the fabric hands which express fabric character. These hands are classified by several expressions, each of which relates to the fabric's mechanical performance regarding clothing and comfort, such as stiffness (KOSHI), smoothness (NUMERI) etc. These hands were named primary hands. These hand expressions can be terms such as "feel strong", "a little weak" etc. It was found during the survey of these hands that, in spite of the fact that the experts had not previously discussed fabric hand with each other, they had used the same classification of primary hands. Another finding was that the feeling intensity of the primary hands could be expressed numerically, such as 10, 9, 8, . . ., in order of intensity.

The standardization of primary hands was begun by the HESC committee according to these findings. The first step was the selection of the important primary hands for the classification of each fabric and the definition of each primary hand. Table 7.1 shows the primary hands of men's suitings and ladies' thin dress fabrics. For the latter category, two more were added as also shown in Table 7.1. The definition of these primary hands are shown in Table 7.2.

Table 7.2 Definition of the primary hands

Handle (Japanese term):	(English equivalent)	Definition
KOSHI	Stiffness	A feeling related mainly to bending stiffness. A springy property promotes this feeling. A fabric having a compact weave density and made from springy and elastic yarn gives a high value
NUMERI	Smoothness	A mixed feeling coming from a combination of smooth, supple, and soft feelings. A fabric woven from a cashmere fiber gives a high value
FUKURAMI	Fullness & softness	A feeling coming from a combination of bulky, rich and well formed impressions. A springy property in compression and thickness, accompanied by a warm feeling, is closely related with this property. (The Japanese word literally means „swelling")
SHARI	Crispness	A feeling coming from a crisp and rigid fabric surface. This is found in a highly woven fabric made from a hard and strongly twisted yarn. This gives a cool feeling. (The Japanese word means crisp, dry and sharp sound caused by rubbing the fabric surface with itself)
HARI	Anti-drape stiffness	The opposite of limp conformation whether the fabric is springy or not. (The Japanese word means "spread")
SHINAYAKASA[a]	Flexibility with soft feeling	Soft, flexible and smooth feeling
SOFUTOSA[a]	Soft feeling	Soft feeling, a mixed bulky, flexible and smooth feeling

[a] The SHINAYAKASA and SOFUTOSA are not primary hand but semi-total hand

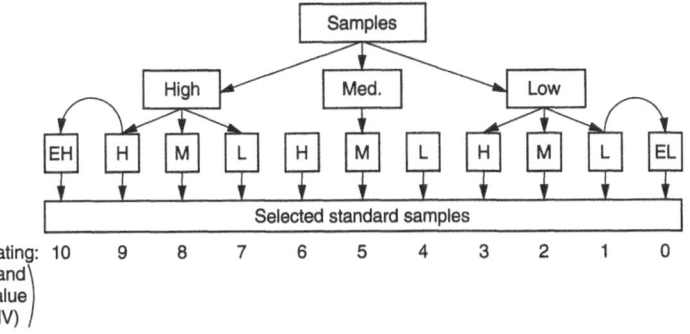

Fig. 7.4 The grading of hand value of the primary hand and selection of the standard samples. This procedure was repeated for each primary hand

The second step was to select the standard samples for each of the primary hands and about 600 commercial fabric samples were collected. Grading was carried out for each primary hand as shown in Figure 7.4 by each expert individually. The assessment was done separately for every primary hand by the experts. A judge touched a fabric and if he felt a strong or weak hand, he put the fabric in the strong feeling group or weak feeling group respectively. If he could not decide, he put it in the medium feeling group. The setting of this fuzzy zone made the assessment so easy that the experts could complete the judgement in a short period, approximately half a day for all 600 samples. All samples were separated into three groups. Then the same procedure of the judgement was repeated for each of the three groups, then all samples were separated into nine groups. Those samples having especially strong or weak feeling were again separated from the highest and the lowest group of the nine groups respectively as shown in Figure 7.4. After all the experts had judged all the samples for every primary hand, those samples which were graded unanimously were selected to define the standard and numbered from 10 to 1 accordingly. These numbers are known as "Hand value" or "H. V.".

Each primary hand was then standardized and copies of the standard samples were published for men's winter and summer suiting in 1975 [7.4] and ladies' thin dress

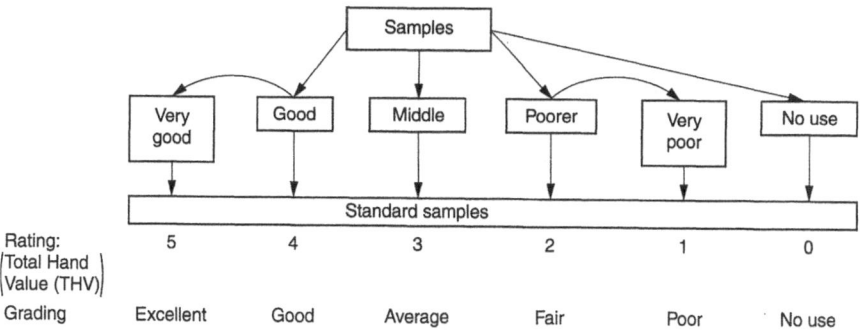

Fig. 7.5 The grading of the total hand and selection of the standard samples for each grade

Table 7.3 Inter-relation between the pirmary hands and fabric function as clothing materials

Stiffness (KOSHI): makes a moderate space between human body and clothing, good shape retention and beautiful silhouette of garment

Smoothness (NUMERI): increases smooth-touch feeling and soft-touch feeling

Fullness (FUKURAMI): increases space between fibers and fiber movement; this makes fabrics more extensible and soft

fabrics in 1980 [7.5]. Following this work, the next quality standardization was carried out. An overall hand is expressed as "good hand" or "poor hand". This overall hand expresses fabric quality. The request to the experts was to judge only "good" or "poor". They graded the fabrics of 214 samples selected for this test for commercial winter suitings and 156 for commercial summer suitings following the procedure as shown in Figure 7.5. The procedure for the standardization was similar to that of primary hands standardization except the number of grades. The numbers which were put in order of high grade were from 5 to 1 (0 was considered to be not suitable), and named "Total hand value" or "T.H.V.".

Standardization of the Total hand was carried out for men's winter and summer suiting, ladies' thin dress fabrics and some other classified fabric groups such as men's jacket, men's dress-shirt fabrics etc. This subdivided classification was necessary in the case of T.H.V. as a quality hand.

The standard samples of T.H.V. were published in 1983 [7.6] only for men's winter suiting and this publication effort is still continuing for men's summer suiting and ladies' thin dress fabrics.

T.H.V. is judged considering the hand values of primary hands. The judgements of both the primary hands and the total hand are done subjectively by each expert on the basis of their criteria. Interestingly, the difference between the criteria of the experts is not great for either the primary hand or total hand judgements. This is because the experts have based their subjective criteria on experience and knowledge gained from customer feedback. It is noted that the primary hand such as stiffness (KOSHI) etc. is not purely emotional feelings, but closely related to fabric function as clothing material. Table 7.3 shows the inter-relation between the primary hand and the fabric functions.

7.4
Objective Evaluation of Fabric Hand

Parallel to the standardization of the hands, preparations for the objective evaluation of the hands were carried out. The concept of this objective evaluation is shown in Figure 7.6.

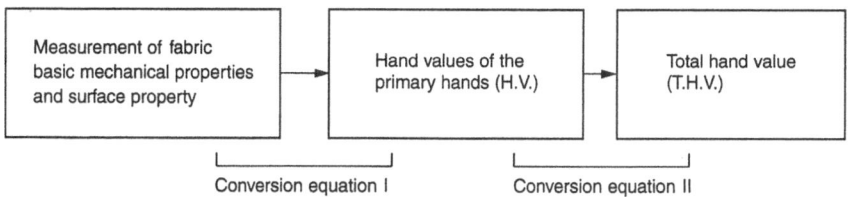

Fig. 7.6 The sequence of the objective evaluation

The first stage of activity of HESC was the preparation of the measuring facilities to determine mechanical and surface properties of fabric. The mechanical parameters selected for the objective evaluation were, firstly, that it had to be well correlated with the subjective evaluation and, secondly, that parameters had to be measured under the basic deformation of fabric because of the future connection of the hand with fabric, yarn or fiber design of high quality fabric.

Under the conditions shown above, a measuring system was designed by Kawabata based on his basic research on fabric mechanical properties which had been carried out by his group. This system consisted of four instruments, these being the tensile and shearing tester, pure bending tester, compression tester, and surface smoothness tester. These testers easily measured the fabric mechanical properties in the low-load region and provided a large quantity of data for many samples which were inspected for fabric hands – using the subjective method – by the experts in HESC.

Figure 7.7. shows the bending tester for fabric. A sample 1 cm in length and 20 cm in width is bent by two chucks. One of these chucks moves on a locus which provides pure bending of fabric and the other detects the bending torque as a function of curvature. Figure 7.8 shows an example of the bendingt curve of a worsted weave. The slope B and hysteresis of torque 2HB are measured as the mechanical parameters.

Figure 7.9 shows the surface tester measuring the surface friction and geometrical roughness. The surface of the friction detector is made of steel wires having a special

Fig. 7.7 The pure bending tester

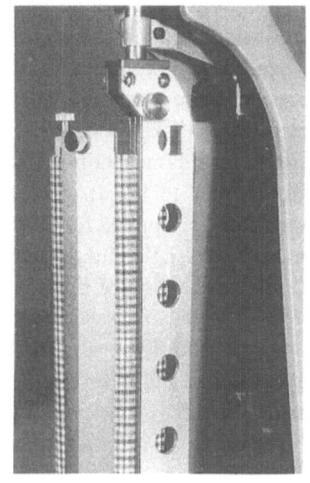

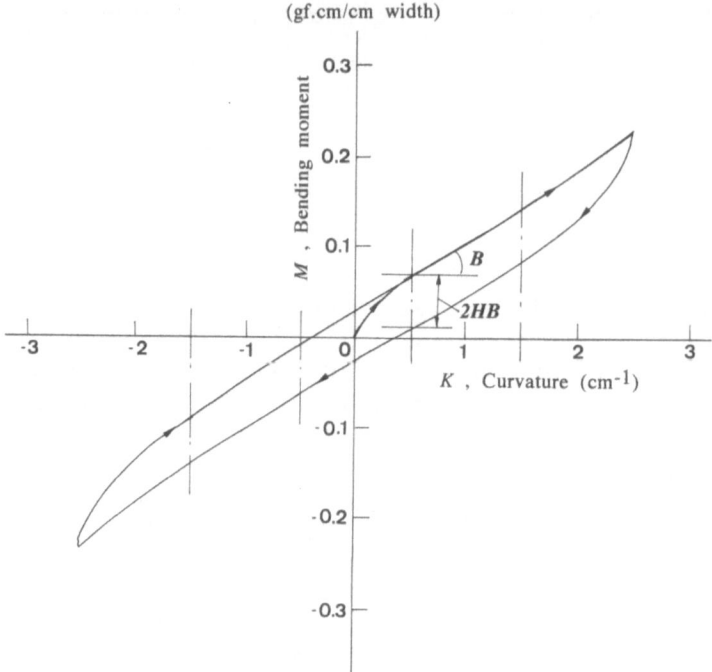

Fig. 7.8 A typical curve of the bending moment-curvature of a worsted fabric

shape as shown in Figure 7.9b. This shape simulates human finger marks and has been developed after preliminary investigation on the human sensitivity of surface roughness. The signal from the contactor goes through the filter having the frequency response shown in Figure 7.10. The data processing by this filter increases the sensitivity of the roughness detection by the contact element. The mechanical parameters used for the objective evaluation of the fabric hand are shown in Table 7.4. All the instruments of the system are shown in Figure 7.11.

On the basis of the data from the subjective evaluation, the equations which could connect these mechanical parameters with the hand values (H. V.) and then the total hand value (T.H.V.) were derived by the present authors by statistical analysis of the mechanical data obtained instrumentally and the hand values judged subjectively by the experts. In the H. V. derivation for each primary hand, a multivariable linear equation was applied for H. V. and a non-linear multivariable equation having square terms for T.H.V. The detail of the conversion equations are as follows.

The conversion of the mechanical parameters into primary hand values is

$$Y_i = C_{i0} + \sum_{j=1}^{16} C_{ij} x_j \tag{7.1}$$

or

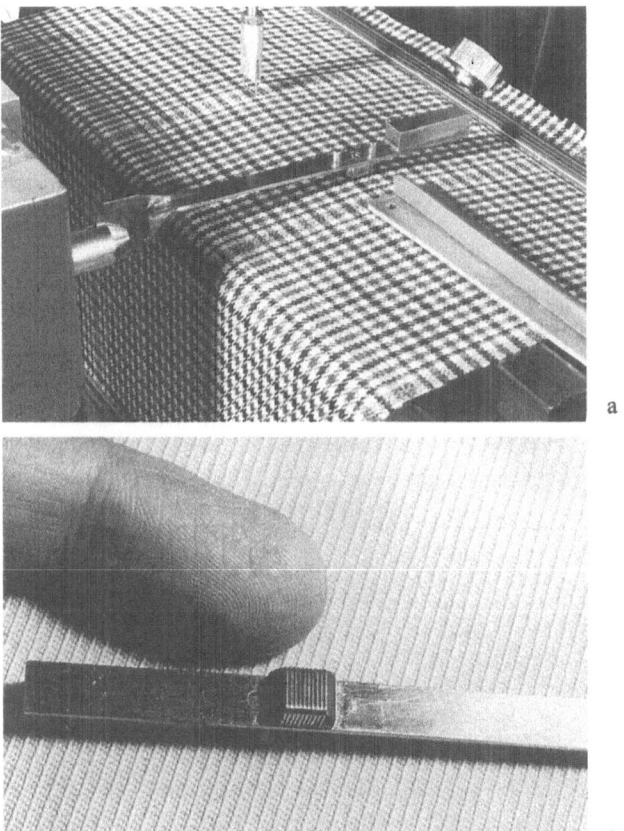

a

b

Fig. 7.9 a The surface roughness tester. Surface geometrical and frictional roughness are measured by the two separated contactors. b The contactor used for the surface frictional roughness. The finger mark (top) is simulated by thin steel wires

$$
\begin{bmatrix} Y_1 \ (Stiff) \\ Y_2 \ (Smooth) \\ Y_3 \ (Full) \end{bmatrix} = \begin{bmatrix} C_{1,0} & C_{1,1} & C_{1,2} \cdots C_{1,16} \\ C_{2,0} & C_{2,1} & C_{2,2} \cdots C_{2,16} \\ C_{3,0} & C_{3,1} & C_{3,2} \cdots C_{3,16} \end{bmatrix} \begin{bmatrix} 1 \\ x_1 \\ x_2 \\ \cdot \\ \cdot \\ \cdot \\ x_{16} \end{bmatrix} \tag{7.1$'$}
$$

This equation is for men's suiting where

Y_i: Hand value of i-th primary hand, $i=1,2$ and 3,

C_{i0}: Constant,

C_{ij}: Constant coefficient, $i = 1,2,3, j = 1,2,\ldots, 16$,

xj: j-th normalized mechanical parameter. This parameter is derived by normalizing parameter as follows.

Table 7.4 The mechanical pa- rameters used for the objecti- ve evaluation			
TENSILE	1.	LT	Linearity
	2.	WT	Energy
	2.	EM	Extensibility
	3.	RT	Resilience
SHEAR	4.	G	Rigidity
	5.	2HG	Hysteresis
	6.	2HG5	Hysteresis (High shear)
BENDING	7.	B	Rigidity
	8.	2HB	Hysteresis
COMPRESSION	9.	LC	Linearity
	10.	WC	Energy
	11.	RC	Resilience
SURFACE	12.	MIU	Coefficient of friction
	13.	MMD	Frictional roughness
	14.	SMD	Geometrical roughness
CONSTRUCTION	15.	W	Fabric weight
	16.	T0	Fabric thickness

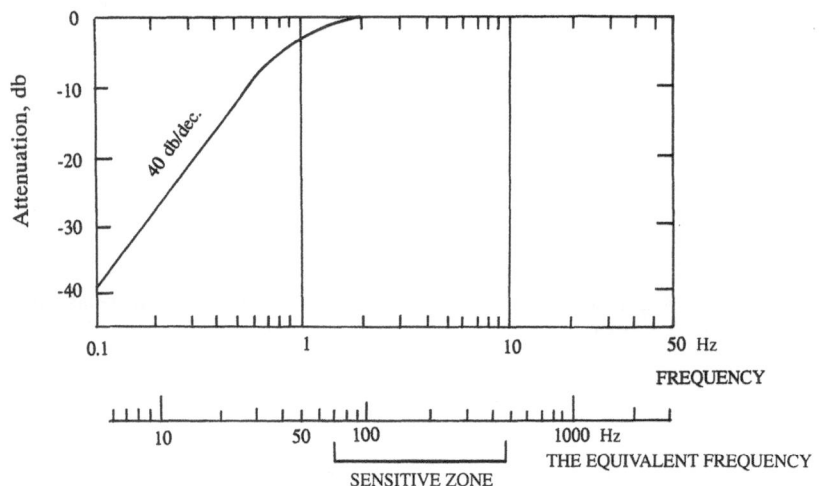

Fig. 7.10 The frequency response of the filter used for data processing for the surface roughness signals. The sweep velocity in the measurement is 1 mm / sec. Frequency range 1 ~ 10 HZ corresponds to 50 ~ 500 HZ in actual sweep velocity of finger. And human finger is the most sensitive in this frequency range. The filter eliminates the lower frequency component

$$x_j = (X_j - M_j)/\sigma_j \tag{7.2}$$

where X_j is j-th mechanical-parameter, $j = 1,2,3, \ldots, 16$. Some of parameters are the transformed form by logarithm [1],

M_j: mean of j-th mechanical parameter, X_j,

σ_j: Standard deviation of j-th mechanical parameter, X_j,

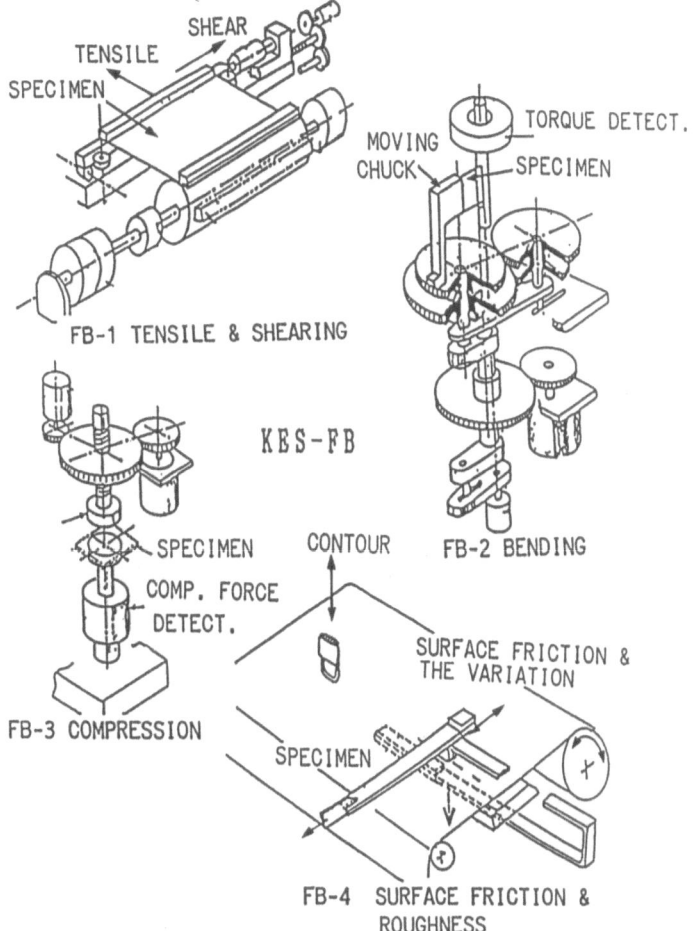

SHEAR

TENSILE

SPECIMEN

FB-1 TENSILE & SHEARING

MOVING
CHUCK

TORQUE DETECT.

SPECIMEN

KBS-FB

CONTOUR

FB-2 BENDING

SPECIMEN

COMP. FORCE
DETECT.

FB-3 COMPRESSION

SPECIMEN

SURFACE FRICTION &
THE VARIATION

FB-4   SURFACE FRICTION &
ROUGHNESS

Fig. 7.11 The fabric mechanical testing system which is known as the KESF system

The primary hand values Y_i (i=1,2,3) are then converted into T.H.V. as follows

$$\text{T.H.V.} = C_{00} + \sum_{i=1}^{3} Z_i$$

where

$$Z_i = \begin{bmatrix} C_{i1} & C_{i2} \end{bmatrix} \begin{bmatrix} Y_i \\ Y_i{}^2 \end{bmatrix} \tag{7.3}$$

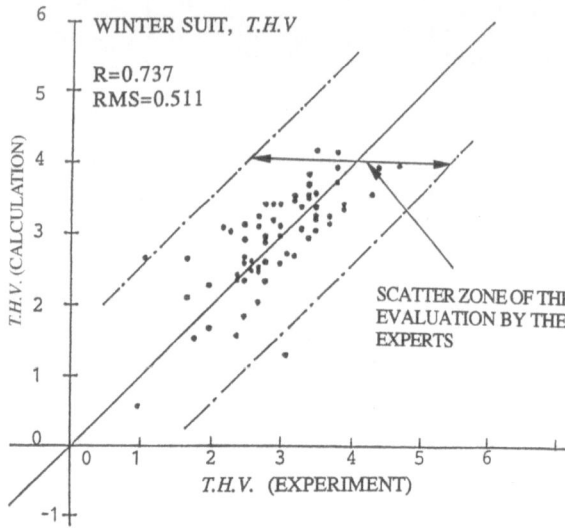

Fig. 7.12 Correlation between the average value of T.H.V. evaluated by 12 experts and the objectively evaluated values

WINTER SUIT, *T.H.V*

R=0.737
RMS=0.511

T.H.V. (CALCULATION)

SCATTER ZONE OF THE EVALUATION BY THE EXPERTS

T.H.V. (EXPERIMENT)

where
Y_i: H. V. of i-th primary hand
C_{00}:   constant
C_{i1}, C_{i2}: Constant coefficient.

By the two step conversion, the mechanical parameters are transformed into T.H.V. as follows:

$$[X_i] \xrightarrow{\;[C_{ij}]\;} [Y_j] \xrightarrow{\;[C_j]\;} T.\,H.\,V. \tag{7.4}$$

Mechanical   Primary
parameters   H. V.
$i=1,...,16,$ $j=1,...,3$ (or 4 for summer suiting)

These equations can evaluate the fabric hand values and total hand values well, and estimate the values near to the average of the scattered hand values evaluated by different experts. Although the scattering of the experts is relatively large, particulary in the evaluation of T.H.V., the objective equation can estimate well the average of the

Table 7.5 Prediction ability of the objective evaluation of T.H.V. This prediction ability was surveyed by 12 experts with 154 samples of men's winter suiting

	Standard deviation
Objective measurement	
Standard deviation of prediction error[a]	0.55
Subjective measurement	
Standard deviation of T.H.V. judged by 12 experts	0.67

[a] Deviation from mean of 12 experts' judgements

scattering as shown in Figure 7.12. Details of the prediction ability of the equations is shown in Table 7.5. This high capability of the objective system in the hand evaluation is the basis for the widespread usage of this objective system in the fiber and textile industries.

7.5
Developments in the Application of Objective Evaluation

After the establishment of the basic technology for the objective evaluation of fabric hand, application of this technology has spread rapidly in the fiber and textile industries and other industries producing "human materials".

As described before, the fabric mechanical properties under a low-load region are applied to the objective hand evaluation. Around 1975, it was found by a HESC-subgroup that some of these mechanical data correlated well with fabric processability [7.1, 7.7] in clothing manufacture and also with suit making-up [7.1, 7.8, 7.9]. This processability had also been examined subjectively by expert hand inspection. This new application augmented the application of objective measurement. There are two major applications at present, as follows.

(1) The objective evaluation of fabric hand and application of the hand evaluation to other materials such as leather, paper, cosmetics and other "human materials".
(2) Use of mechanical data for the process control in clothing manufacturing and predicting the appearance of a suit.

Some recent applications in categories (1) and (2) are introduced here.

7.5.1
Fabric Characterization

Table 7.6 shows an example the characterization of men's suiting by means of the objective method. A simple but useful chart is also showed in Figure 7.13 where the data on Table 7.6 is plotted. This chart shows the clear target to develop new suitings. The shadowed zone is the zone of particularly high grade T.H.V. suitings, a clear target to develop the new suitings. This data table is powerful, the fabric hand now being characterized without the experts' subjective judgement. The numerical expression of the primary and total hand is especially important and useful for the trace of process control to see the effect of actions applied to the fabric. Figure 7.14 also shows the hand on a snake curve, useful to see the feature of fabric character in terms of

Table 7.6 An example of the characterization of fabric hand

Winter suiting sample	'1	'2	'3	remarks
T.H.V.	4.5	3.7	2.7	Max. value: 5
H.V.				
Stiffness (KOSHI)	4.0	3.6	7.7	Max. value: 10
Smoothness (NUMERI)	7.7	6.7	3.7	
Fullness (FIKURAMI)	7.0	5.3	4.1	

hand; e. g., we can draw a zone of the hand of silk weaves and check whether a fabric is silk-like or not by plotting the hand values on the chart.

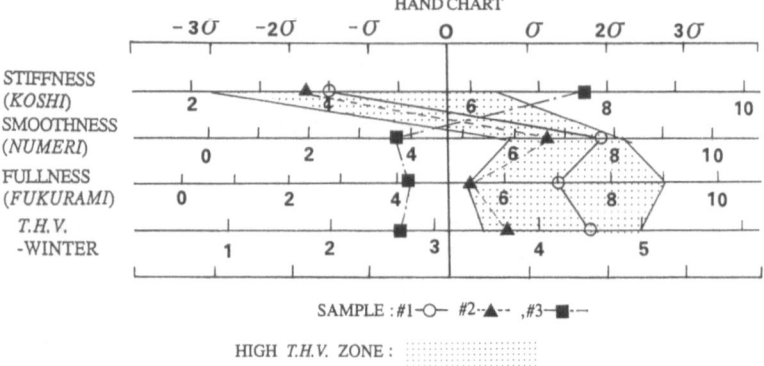

Fig. 7.13 The hand chart for men's winter suiting fabrics. The H.V. and T.H.V. of the three samples in Table 7.6 are plotted in this chart. **The shaded zone** is the excellent zone in which high quality fabrics fall

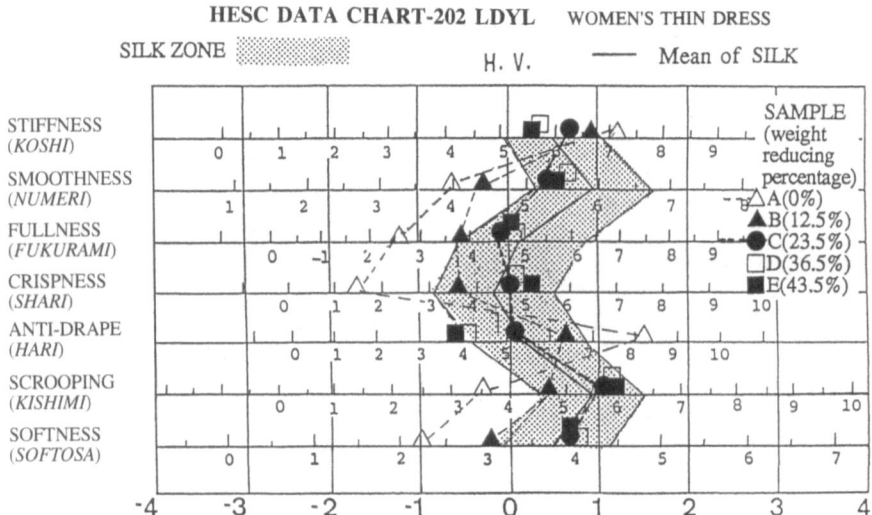

Fig. 7.14 The change of fabric hand of a PET filament weave is inspected on the hand chart for women's thin dress fabrics. With an increasing degree of weight reduction by alkali treatment the fabric approaches the silk hand zone which is shown by **the shaded snake-zone**

7.5.2

Discrimination of Fabric Hand for Fabric Design

Discrimination analysis is useful in the characterization of fabric hand and the design and development of new fabrics [7.3]. We can detect subjectively by hand touching whether a fabric is a silk fabric or a cotton fabric. This is done by our experience on the basis of the feature in the fabric hand. The numerical hand expression enables this discrimination to be made objectively. In a case where there are three groups of fabric, say, cotton, silk, wool fabrics, they can be discriminated by the discriminate variables Z_1 and Z_2 which are derived from their primary hand values by Eq. (7.5):

$$\begin{bmatrix} Z_1 \\ Z_2 \end{bmatrix} = \begin{bmatrix} C_{11}, & C_{12}, & C_{13}, \ldots C_{1n} \\ C_{21}, & C_{22}, & C_{23}, \ldots C_{2n} \end{bmatrix} \begin{bmatrix} Y \\ Y_2 \\ Y_3 \\ \\ Y_n \end{bmatrix} \qquad (7.5)$$

where Y_i ($i = 1 \sim n$) is the hand value of the i-th primary hand and C_{ij} ($i = 1,2, j = 1 \sim n$) are constant coefficients which are derived by statistical discrimination analysis which discriminates three groups of fabric types in terms of fabric hand.

Figure 7.15 shows the discriminated zones of three types of women's thin dress fabrics made of silk, cotton and wool respectively in terms of their hand. For the category of women's thin dress fabric, four more primary hands have been defined in addition to three for men's suiting as shown in Table 1. The discrimination equations are as follows.

$$Z_1 = 6.659 - 1.588Y_1 - 0.952\,Y_2 - 0.511Y_3 + 0.627Y_4$$
$$- 0.296Y_5 + 1.550Y_6 + 0.367Y_7$$

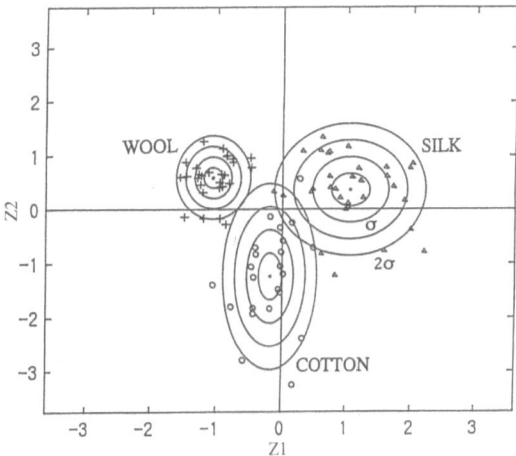

Fig. 7.15 Discrimination of fabric hand by Z_1, Z_2 parameters. Wool-like, cotton-like and silk-like hands are clearly separated by the three zones. **The central area of the zone** is the zone of typical hand.

Fig. 7.16 The change of the fabric hand by the weight reduction treatment shown in Fig. 7.14 is again shown on the discriminating chart. We call this method of hand monitoring "navigation method"

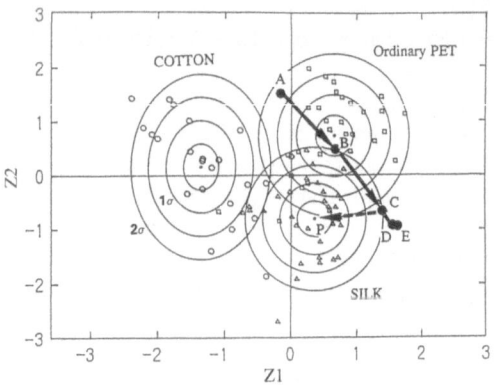

Fig. 7.17a,b Fabric hand of the new synthetic-fibre fabrics (shin-gosen): **a** "New worsted" type; **b** "Peach skin" type

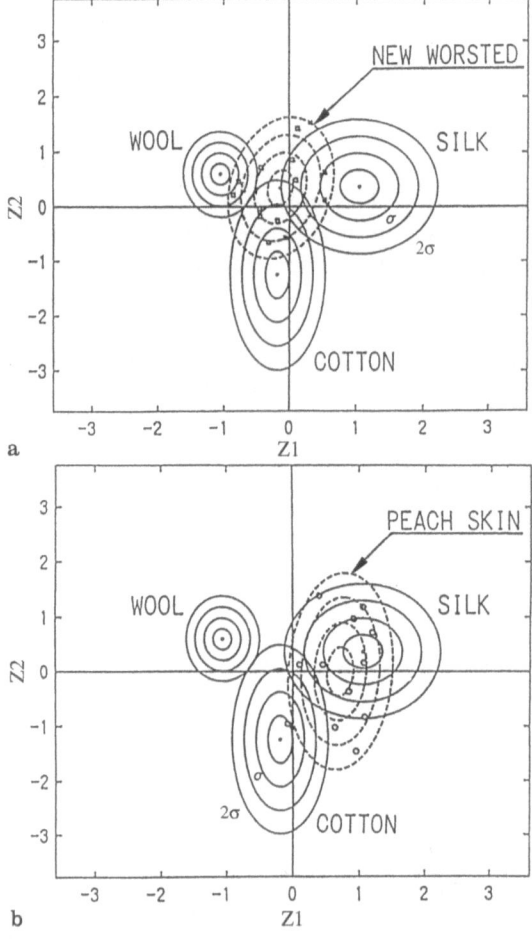

$$Z_2 = -3.111 + 1.521Y_1 - 0.557Y_2 + 0.249Y_3 - 1.131Y_4 \qquad (7.6)$$
$$- 0.018Y_5 - 0.079Y_6 + 0.815Y_7.$$

They are separated by the three distribution zones on Z_1 and Z_2 domain. Each zone is shown by contour lines of standard deviations $\sigma = 0$ (central position), $\sigma = 0.5$, $\sigma = 1.0$, $\sigma = 1.5$ and $\sigma = 2.0$. If the set of Z_1 and Z_2 of a fabric sample falls in, for example, the silk zone, the fabric hand of this sample is silk-like and when the point is near the center of the distribution, the hand of that fabric specimen is a typical silk type.

Figure 7.16 shows the change in hand of a PET filament weave sample by alkali dehydration weight-reduced treatment shown on the cotton/silk/PET discrimination chart. With an increasing percentage of the weight of the fabric reduced, the hand approaches the silk zone. However, it is shown clearly that over-treatment (C,D,E) moves the fabric hand very far from silk. To obtain a more silk-like PET fabric, its direction must be corrected. We named this the "Navigation" method, useful for new fabric development.

Some plotted data [7.10] on the Z_1–Z_2 map in Figure 7.17 are examples of new fabrics called "shin-gosen". Shin-gosen means "new synthetic-fiber fabric". These fabrics have been developed rapidly in the past few years in Japan and are now very popular. The important concept of this fabric is its new hand character designed by the modification of mainly PET fiber, such as mixing of irregular denier fibers, use of micro-denier fibers etc.

This discrimination shows two types of shin-gosen fabrics, the "new worsted" type and the "peach skin" type. The hand character of these two types are easily compared with the hand to the three traditional fabrics – silk, cotton and wool fabrics.

7.5.3
Development of New Fabrics

Figure 7.18 shows how we can develop new high quality fabrics [7.1, 7.11]. Joint research by HESC and the New Zealand Wool Research Organization (WRONZ) was initiated in 1985. The aim of the research has been to develop a high quality summer suiting by replacing traditional mohair wool with New Zealand coarse wool of 35 μm diameter. In this figure, a mechanical parameter chart and a hand value chart are shown. The snake zone seen in the chart is the zone in which high quality summer suitings fall, this zone being derived by the survey carried out by HESC. The filled symbols are the properties of high quality and mohair-containing summer suitings. The snake zone and snake line of target fabrics, mohair-containing fabric in this case, are the guidelines for the new fabric development. The experience that experts of fabric design have gained in the objective method has initiated the rapid development of high quality fabric using the New Zealand coarse fiber.

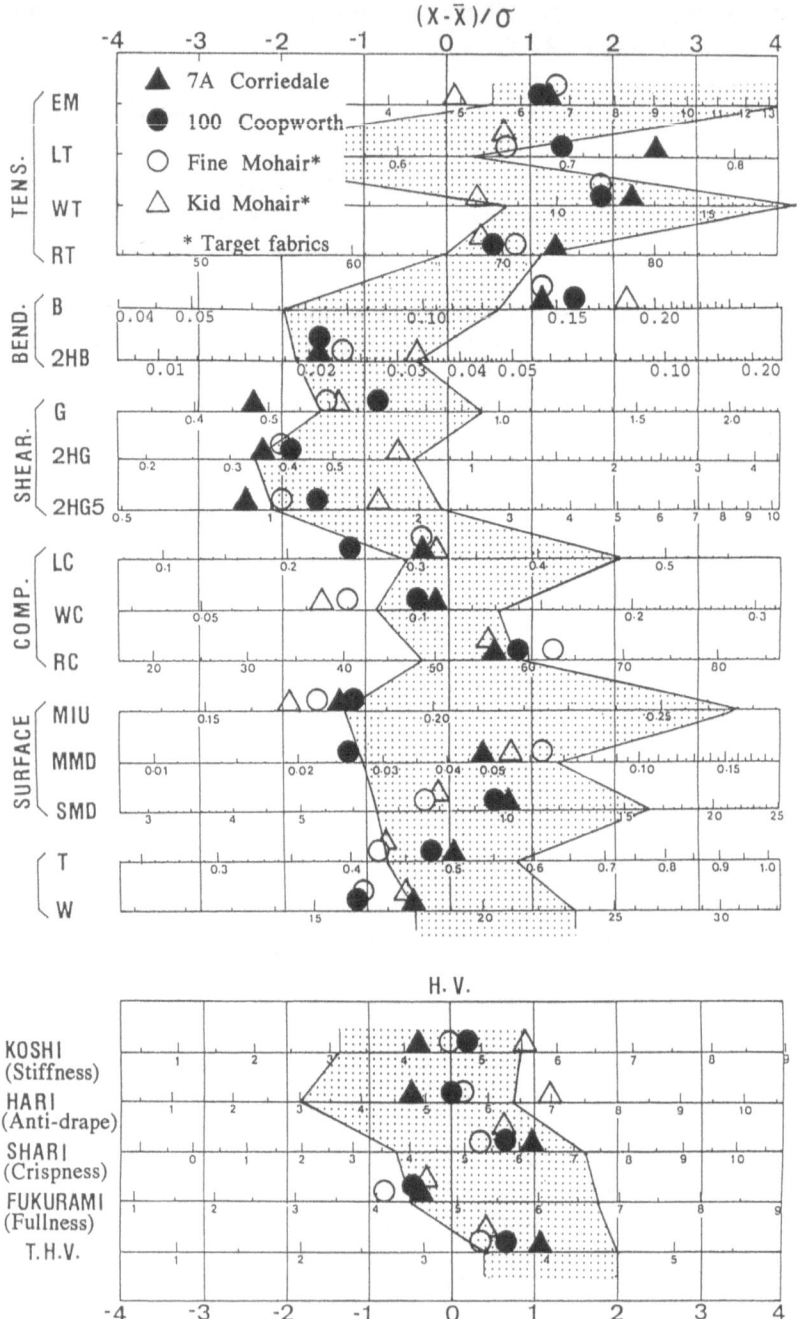

Fig. 7.18 Development of new high-quality fabrics. The property of the target fabrics is that of the men's summer suiting made of mohair-containing fabrics. **The filled symbol** is the fabric property which is approaching the property of the target fabrics

7.5.4
Quality of Artificial Leather

It is interesting that the primary hand and T.H.V. which have been developed for men's suiting can also predict the quality of other human materials, examples of which are introduced here.

Leather is also a typical "human material". It is noted that the quality with which we are concerned here is that coming from hand touch. The primary hand and T.H.V. of artificial leather materials used for clothing such as coats and jackets were calculated based on the equations developed for men's suiting.

The mechanical parameters X_i are normalized by the mean value m and standard deviation σ of the leather material group [7.13]. The normalized parameters are then substituted into the primary hand equation developed for men's suiting. The T.H.V. were obtained by exactly the same equation as the men's suiting equation.

The prediction ability was examined by using two groups of artificial leather samples. One is a group of excellent quality samples and the other a group of poor quality samples. These samples were subjectively evaluated by experts in artificial leather manufacturing factories. The experts evaluated these samples in the same manner as the T.H.V. evaluation of men's suiting, i. e., by the five grading method. The excellent samples are those with a T.H.V. higher than 4.0 and the poor samples those with a T.H.V. lower than 2.0.

The objective values of the primary hand values and the T.H.V. of these two groups are plotted on the hand and quality chart used for men's suiting as shown in Figure 7.19. It is shown that the excellent samples, plotted by circles, are clearly separated from the poor samples, plotted by crosses, by NUMERI (smoothness), FUKURAMI (fullness), SOFUTOSA (softness) and also by T.H.V. Here, SOFUTOSA (softness) is not a primary hand but a semitotal hand expressing the soft feeling of women's suitings.

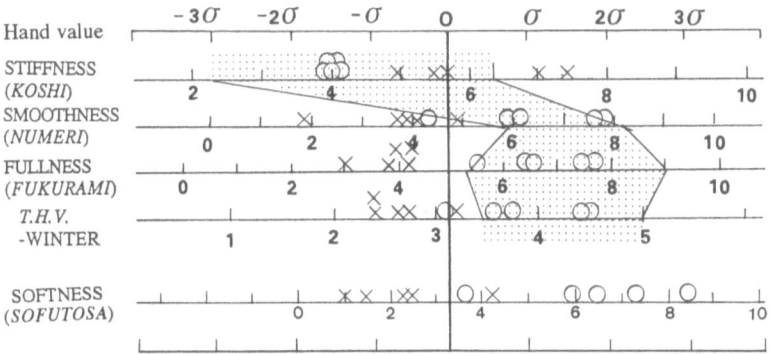

Fig. 7.19 The hands of artificial leather are plotted on the hand chart for men's winter suitings. The hands of excellent samples (symbol o) fall in the same good-zone as the men's suiting and the poor samples (symbol x) fall outside the zone

It is also interesting that the excellent samples fall in the snake zone which is the high quality zone of men's suiting, and the poor samples fall outside the good zone.

7.5.5
Quality of Facial Tissue Paper

Investigations similar to those for artificial leathers has been carried out for facial tissue papers. The H.V. and T.H.V. of three typical samples are plotted on the hand and quality inspection chart for men's suiting as shown in Figure 7.20. The hand values of the good paper fall in the good zone and the poor paper is outside the zone. The average quality paper is closer to the good zone but not inside it. The subjective judgement of paper quality is based on the assessment which was carried out by 10 specialists working in hairdresser's shops and 10 consumers.

The assessment by the specialists showed good correlation with the objective evaluation. The correlation coefficient between the subjective T.H.V. and the objective values of each primary hand and T.H.V. are shown in Table 7.7. It is noted that the primary hands are highly correlated with the subjective quality.

The smoothness (NUMERI) is highly correlated with the subjective quality T.H.V. This primary hand is the most important hand for the quality of men's suiting for winter use. The objective T.H.V. is, however, not so highly correlated with the subjective T.H.V. This is because of the difference in the usage between paper and men's suiting.

This result suggests to us that the equation for predicting T.H.V. must be derived for each material, i. e., for the tissue paper group in this case. In case of paper, the importance of smoothness (NUMERI) may be much higher than that in the case of men's suiting.

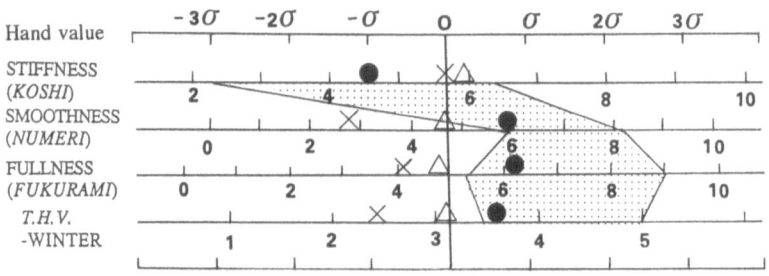

Fig. 7.20 The hands of facial tissue papers are plotted on the same chart as Fig. 7.19. Excellent paper (symbol ●) falls in the good-zone. Average and poor samples (symbol Δ and X respectively) fall out the good zone.

Table 7.7 The correlation coefficient between subjective T.H.V. and predicted hand values/T.H.V. of facial tissue papers. The subjective values were evaluated by professionals in beauty salons. The objective values were derived by the equations for men's winter suitings		Correlation coefficient
	Stiffness (KOSHI)	– 0.24
	Smoothness (NUMERI)	0.95
	Fullness (FUKURAMI)	0.85
	T.H.V.-Winter suiting	0.84

7.5.6
Quality of Automobile Upholstery

The quality of the sheet material for covering the dashboard and front panel of automobiles was examined using the objective evaluation. This material, used as the covering sheet of the dashboard and steering wheel, is made of polymers. The sheet, as face material, is laminated with the body material. We call this face material "panel sheet", and the quality of this panel sheet has been judged by the experts in panel sheet manufacturing factories.

Judgement was made by the hand of the sheet itself, not in its laminated state, because there is a high correlation between the hand of the sheet and the composite panel body. The evaluation of the subjective T.H.V. was carried out in the same manner as the T.H.V. assessment of men's suiting.

The three groups of the panel sheet were sampled as follows:

excellent group: S.T.H.V. > 4.0

average group: 3.5 > S.T.H.V. > 2.5

poor group: S.T.H.V. < 2.0

where S.T.H.V. is subjective T.H.V.

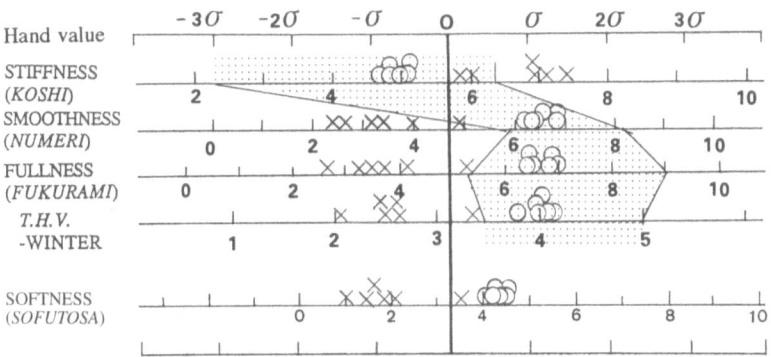

Fig. 7.21 The hands of the panel sheet of automobiles are plotted on the same chart as Fig. 7.19. Excellent sheet (**symbol** O) falls in the good-zone of men's suiting, poor sheets (**symbol** X) are completely outside.

The objective H.V. and T.H.V. of the excellent and the poor groups are plotted on the hand and quality inspection table for men's suiting as shown in Figure 7.21. In spite of quite different material from the men's suiting, th H.V. and T.H.V. of the excellent group (circle symbols) fall in the good zone of men's suiting, and the poor group is completely outside the good zone. The objective H.V. and T.H.V. were also calculated using the equation for the H.V. and T.H.V. evaluation of men's suiting in the same manner as in the case of leather materials.

These three examples suggest to us that there is a general criterion by which humans feel "good" hand in all human materials even though the details of the criteria for each material is slightly different because of the difference of its function in use.

7.6
Conclusion

The objective method for fabric hand evaluation has now spread throughout the fiber, textile and clothing manufacturing industries. One of the recent trends in these industries is the shifting towards producing high quality fabrics. This trend is natural because human life is undoubtedly going up to a higher level than in the past. The twenty-first century will put much value upon human life, and clothing will play an important role in the improvement of human life in the future.

This chapter is based on the paper presented by the present authors at the International Conference "Textile Science 91" held at the Technical University of Liberec, Czech Republic in September 1991 [7.12].

7.7
References

7.1   Kawabata S and Niwa M (1989) Fabric performance in clothing and clothing manufacture, J Text Inst 80: 19
7.2   Peirce F T (1930) The 'Handle' of cloth as a measurable quality. J Text Inst 21: T377
7.3   Kawabata S (1980) The standardization and analysis of hand evaluation, 2nd edn HESC, Textile Machinery Soc Japan, Osaka
7.4   Kawabata S (ed) (1975) HESC standard of hand evaluation (HV standard for men's suitings). HESC, Textile Machinery Soc Japan, Osaka
7.5   Kawabata S (ed) (1980) HESC standard of hand evaluation, 2nd edn. vol 1 (HV standard for men's suitings); Vol 2 (HV standard for women's thin dress fabric). HESC, Textile Machinery Soc Japan, Osaka
7.6   Kawabata S (ed) (1983) HESC standard of hand evaluation, Vol 3 (THV standard for men's winter suitings). HESC, Textile Machinery Soc Japan, Osaka
7.7   Ito K and Kawabata S (1985) Conception of the automated tailoring controlled by fabric objective-measurement data in "Proceedings of 3rd Japan / Australia Symposium on objective measurement: applications to product design and process control" (edited by Kawabata S, Postle R and Niwa M), Textile Machinery Soc Japan, Osaka, p 175
7.8   Niwa M, Yamada T and Kawabata S (1981) Prediction of the appearance of men's suit from fabric mechanical properties and fabric hand, J Textile Machinery Soc Japan, 34, T12, T76, T135

7.9 Niwa M and Kawabata S (1988) The three mechanical components of fabric relating to suit appearance in application of mathematics and physics in the wool industry, WRONZ, p 404

7.10 Koyama Y, Niwa M and Kawabata S (1991) An analysis of the hand of shingosen weaves, Proc of the 20th Text. Res Symposium, the Text. Machinery Soc Japan

7.11 Kawabata S, Carnaby G A and Niwa M (1988) New Zealand/Japan joint project for developing high quality summer suitings using New Zealand wool and fabric objective measurement technology in application of mathematics and physics in the wool industry, WRONZ, p 92

7.12 Kawabata S and Niwa M (1991) Recent progress in the objective measurement of fabric hand, Proc of The International Conference "Textile Science 91", Technical University of Liberec, Czechoslovakia

7.13 Kwabata S, Niwa M and Wang F (1994) Objective hand measurement of nonwovens fabrics, part I: Development of the eguation, Textile Research Journal, vol 64, p 597

Chapter 8

Impact of Medical Technology Utilizing Macromolecules on Society

Yutaka Sakurada and Koichi Takakura

Abstract

This chapter, covering medical applications, consists of three sections: overview, hollow fiber application and dental application. The first section is an overview of the wide range of utilizations of macromolecules with medical devices, including artficial organs and various other hospital products. The reasons why macromolecules are used so widely with medical devices and sometimes drugs are discussed, together with recent progress with polymeric biomaterials with regards to their application to medical devices. Such medical devices, in which macromolecules play such a key role, and the impact of new medical technology born from those medical devices on society are identified. The second section deals with hollow fiber applications, including details of recent developments in blood purification devices using hollow fiber membranes made from macromolecules. This section has been prepared from the contribution of Dr. Sueoka. The third section emphasizes the impact of the successful development of polymeric adhesives on dental materials and dentistry. This section has been prepared from the contribution of Dr. J. Yamauchi. The authors, Dr. Sakurada and Dr. Takakura, had been working together with Dr. Sueoka and Dr. Yamauchi in the Medical Products Division of Kuraray Co., Ltd., which specialises in medical applications of hollow fiber membranes and dental material.

8.1
Overview

8.1.1
Introduction

Significant progress in medical technology has been achieved by the successful coupling of medical hardware, represented by medical devices, and software, which has been established by clinical practice during the two decades from 1970 to 1990. Comparing the limited ways of giving drugs, such as oral, subcutaneous, arterial and intravenous injection or suppository, which are mostly equivalent even for new drug products, new medical devices always need the development and creation of totally new therapeutic and diagnostic methods. Medical technologies born in this period have made a remarkable contribution to man by providing ways of maintaining life,

Okamura, Rånby, Ito (Eds.): Macromolecular Concept and Strategy
© Springer-Verlag Berlin · Heidelberg 1996

replacing an organ that has lost its function, shortening treatment time, reducing pain during treatment, or earlier and more precise diagnosis. On the other hand, however, sometimes there arise problems such as lack of quality of patient life when new medical technologies are introduced.

Macromolecules are playing an important and sometimes indispensable role for many medical devices, although not medical electronics equipment which have also shown remarkable progress, including a series of new equipment such as that for X-ray CT or magnetic resonance imaging (MRI). Relevant medical devices include artificial organs, numerous hospital products which are used as goods for medical care, as tools for auxiliary processes or diagnosis, special therapeutic devices, and dental materials. A well accepted and most widely used artificial organ is the artificial kidney which is often referred to as hemodialysis, and the intraocular lens. Hemodialysis membrane made from macromolecules by the specially developed formation technology gives artificial kidney for use as an extracorporeal device, which is the essential part of this life-sustaining device for patients who have lost their kidney function. Intraocular lenses, implanted as a lens replacement after cataract operations, are made mainly from polymethyl methacrylate (PMMA), and exhibit function as an optical lens. There are other artificial organs which are clinically available or under different stages of development. Artificial joints, artificial blood vessels (vascular grafts) and artificial valves are used widely as implant for replacement of patients' own organs. Macromolecules are used as main constituents of these artificial organs, utilizing their biocompatibility, including antithrombogenicity, and their mechanical properties. Artificial lung is an extracoporeal device used as oxygenator during open-heart surgery, and also as a long-term lung assist for the treatment of respiratory failure patients. In this device, membranes which are core elements for membrane oxygenation, now clinically recognized as a more acceptable system compared with bubble oxygenation, are based on synthetic polymer materials. The artificial heart, which has been under development as a longstanding, ambitious yet controversial, project, has made considerable progress, even clinically, as total heart replacement, and a heart assist device for shorter periods of time. Macromolecules having anti-thrombogenic surfaces are again playing an important role for the artificial heart. Active research programs have progressed in artificial liver and artificial pancreas. These approaches involve hybrid artificial organs or bioartifical organs which are usually a hybrid between the biological cell and non-biological artificial material, required for immune cut-off between the cell and recipient, and control of cell function. Macromolecules, natural or synthetic, are usually the choice for artificial matrix materials.

Hospital products cover a wide range of applications, routine and sophisticated. Routine hospital products include gauzes, masks, gloves, dressings, syringes, catheters, solution and blood transfusion sets, blood bags and so on. There are special type of catheters which are used for X-ray diagnosis or different types of therapeutic applications. These hospital products are made from plastics, rubbers, films, fabrics and nonwoven fabrics, utilizing different fabrication technologies. A number of extra-corporeal blood purification devices are recently developed products which are used for new therapeutic applications such as direct hemoperfusion, plasmapheresis, immunoadsorption, and leucopheresis, ascites treatment, which utilize hollow fiber membrane, polymer based carrier and antithrombogenic polymer material. These extracorporeal devices have sometimes been classified as artificial organs, because the application technologies have been more or less derived from extracorporeal hemodialysis technology, although each technology is different. Contact lenses used for vision correction or post-cataract operation applications are another synthetic polymer product. Newly developed polymer materials are tremendously successful as materials of soft contact lenses and gas(oxygen)-permeable lenses. Progress in dental materials made from polymers also appears significant, many new polymers having specific functions. A number of new dental technologies, using new materials such as adhesives and composite resins, have been clinically established, and widely applied to patients as general practice.

As another example of contribution of macromolecules to medical technologies, drug delivery systems (DDS) have to be mentioned. DDS in which a drug is delivered to a specific site in the body at a predetermined rate for a definite time has been introduced during these two decades, and has been increasingly accepted clinically. Various polymer materials, including biodegradable polymers, are used as release control matrices or carriers for drugs. DDS has had enormous impact on medicine, providing therapeutic and economic benefits for a wide range of diseases.

Macromolecules have met new material needs, required in conjunction with designing new medical devices leading to new medical technology. A large volume of polymer material has been used for medical devices, from routine and simple to sophisticated and complex. Why are macromolecules used so frequently in medical devices? The reasons would be: 1) advancement of polymer technology which can meet the complex and multiple requirements for each device or, sometimes, drug; 2) nature of polymers which can give multiple functions to each medical product and sufficient physical strength under different environments, in many cases in contact with living tissue, and sometimes as implant; and 3) the low cost of polymer which can restrain the constantly increasing health-care expenditure worldwide, and can justify disposable usage to prevent bacterial and viral infection. Advancement of polymer technology has been seen in molecular, morphological and compounding as well as alloying technologies, which are essential for resin tailoring. Another important advancement is fabrication technology of plastics, rubbers, films, and fibers, including extrusion technology of tubes, injection molding technology of parts and components, and fiber spinning technology of hollow fiber membrane. These technology advancements have enabled the design and production of polymers having a specific function and mechanical strength under required conditions. Of all the functions of polymers used for medical devices, biocompatibility combined with the safety of the material is always required, and other functions such as optical properties, material transport, adhesion, antithrombogenicity and so on should be considered as primary functions affecting the efficiency and performance of each medical device. Advancement of fabrication technology, especially that brought about by the introduction of new improved fabrication equipments, has made a remarkable contribution in respect of reducing cost and broadening applications, by enabling the device itself, or parts of the device, to be made thinner, smaller, and lighter, by facilitating automation, or by reducing multiple processes to a single process for the production of even complex and fine structures.

Establishment of different types of sterilization method, such as steam autoclaving, ethylene oxide gas and gamma-ray irraditation, is extending the acceptance of polymer materials for medical devices. Most medical devices are only used clinically in the sterilized form. Because polymer material can be damaged under certain sterilization conditions, the choice of sterilization method is an important consideration when developing new medical devices. Even if the polymer material showed excellent function and adequate mechanical strength, it could not be employed as a component of a medical device without an effective sterilization method.

8.1.2
Medical Devices in which Macromolecules Play a Key Role

8.1.2.1
Polymeric Biomaterials

A wide variety of synthetic and natural polymers are being used in biomedical applications. They are processed or fabricated to be suitable for use in medical devices or prostheses. These polymeric materials include not only the commodity polymers like polyvinyl chloride, polypropylene, and polyethylene, which are mainly used for high volume applications such as blood bags, tubing, syringes and gauze dressings, but also medical-grade polymers specifically developed for value-added products such as medical devices or artificial organs. Although the volume of these polymeric materials used is small compared with other industrial plastics, they contribute immeasurably to the welfare and health of humans. Here, attention is focused on polymeric biomaterials which are used as part of a device in the treatment or replacement of tissues or organs, while maintaining their intended physical and biological effectiveness.

Medical devices may be roughly classified into two major categories – blood-contacting and tissue-contacting devices. The blood-contacting devices currently available include vascular grafts, heart valves, intra-aortic balloon pumps, artificial hearts, left ventricular assist devices, arterial catheters, central venous catheters, and extracorporeal blood purification devices such as blood oxygenators, hemodialyzers, plasma separators and hemoperfusion cartridges. The hybrid artificial pancreas and liver is also now under development. Tissue-contacting devices are, e. g., bones, joint, tendons, ligaments, breast prostheses, intraocular lenses, artificial skin and dental implants and restorative materials. Many of these devices are now fabricated from commercially available polymers, such as PMMA, polyurethane, polyolefin, polysiloxane, polyamides, polyesters, polytetrafluoroethylene(Teflon), polysulfone, polyacrylonitrile, polyvinyl alcohol(PVA) and its related polymers, cellulose, and collagen, all of which must meet the specific requirements of intended devices. These polymers were not originally developed for medical use. However, for example, PMMA has become widely accepted as a biomaterial with extensive clinical applications in orthopedic and reconstructive surgery, dentistry and ophthalmology, and is now being used for hollow fiber membrane in extracorporeal blood purification device. Cellulose has also been widely used as a membrane in blood-purification applications since the beginning of the development of the hemodialyzer. On the other hand, there have

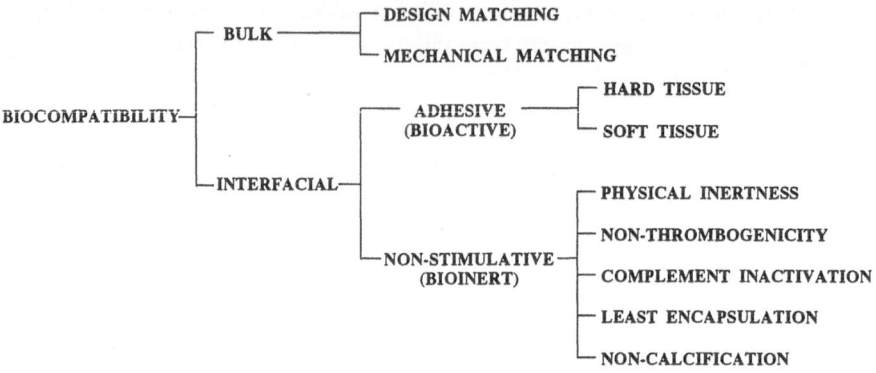

Fig 8.1 Classification of biocompatibility [8.2]

been several examples of tailor-made medical polymers; polyhydroxyethyl methacrylate(PHEMA) originally developed by Wichterle of Czechoslovakia for biomedical application [8.1], thromboresistant segmented polyurethaneurea (Biomer®) and a block copolymer of polydimethyl siloxane and segmented polyurethane (Cardiothane®51, formerly Avcothane®51) developed in the artificial heart program by the National Institutes of Health, U.S.A., medical-grade silicone elastomers for prosthetic devices, and polyglycolic acid or copolymer of glycolic and lactic acids for absorbable sutures.

The polymeric biomaterials come in contact with blood, tissue and organs, and therefore have to meet the requirements for being nontoxic, non-carcinogenic, sterilizable, medico-functional and biocompatible. Their medico-functionality and biocompatibility are important elements for achieving the required high grade of effectiveness and safety of devices. Their medico-functionality covers a wide range of properties such as mass transfer, mechanical, optical, adsorptive and adhesive properties, etc. It must meet the particular design requirement of each device. The term "biocompatibility" has not been well-defined, but the concept proposed by Ikada [8.2] seems to cover all aspects of biocompatibility as represented in Fig. 8.1. In blood-contacting devices, blood compatibility at the blood/material interface is of primary importance. With advances in biomaterials and interdisciplinary research, considerable progress has been made in the development of medical devices. However, with most of the devices, problems still remain in achieving ideal biocompatibility.

8.1.2.2
Blood-Compatible Polymeric Materials

There has been an increasing demand for blood-compatible antithrombogenic materials for blood-contacting applications such as cardiac prostheses, vascular grafts, and extracorporeal blood circulation systems. Currently used biomedical polymers have not proved to be sufficiently antithrombogenic, so that their devices require anticoagulant such as heparin to prevent blood clotting in the device.

Table 8.1 Classification of extracorporeal blood purification therapy with use of membrane and adsorbent

Targeted substances	Principle	Application	Materials* (Membrane or adsorbent)	Diseases
O_2, CO_2	Membrane gas exchange	Membrane oxygenation (MO) / Extracorporeal membrane oxygenation (ECMO)	Polyolefin (PP, PMP) / Silicone, PP-Silicone	Respiratory failure
Low MW	Dialysis	Hemodialysis (HD)	Cellulose, CA, EVAL, PMMA, PAN, PS, PA, PEPA	Renal failure
Middle MW	Filtration	Hemodiafiltration (HDF) / Hemofiltration (HF) / Continuous hemofiltration (CHF)		Hepatic failure
(Albumin)	Adsorption	Direct hemoperfusion (DHP) / Plasma perfusion	Activated carbon / Anion exchange resin	Acute poisoning / Hepatic failure
High MW	Membrane seperation	Plasma seperation (PS) (Plasma exchange) / Plasma fractionation Double filtration Plasmapheresis (DFPP) / Cryofiltration (CF)	PVA, EVAL, CA, / PMMA, PS, PE, PP,	Autoimmune diseases / Metabolic disorders
	Membrane seperation and adsorption	Affinity adsorption or Immunoadsorption	Affinity ligand (biologic and non-biologic) / Carrier beads (Cellulose, PVA, agarose)	Hematologic diseases
Lymphocyte	Adhesion or adsorption	Filtration lymphocytapheresis	Polyester fibers (Non-woven fabrics)	

*CA; Cellulose acetate, EVAL; Ethylene-vinyl alcohol copolymer, PS; Polysulfone, PAN; Polyacrylonitrile, PA; Polyamide, PE; Polyethylene, PP; Polypropylene, PMMA; Polymethyl methacrylate (stereocomplex), PMP; Poly-4-methylpentene-1, PEPA; Polyarylate-polyethersulfone alloy

When blood comes into contact with a foreign surface, a wide range of biological reactions occur – protein adsorption, activation of blood enzymes, complement activation, adhesion and aggregation of platelets, fibrin formation and attraction of white cells. Over the past few decades, extensive studies of biological responses at blood/ material interface have been made by a number of investigators, and nowadays the mechanism of these complex reactions has been elucidated to a considerable extent. The general relationship between biological responses in terms of coagulation (contact phase), complement activation and platelet adhesion, and surface properties of various polymers have been established [8.3], providing valuable information for improving blood compatibility of polymeric materials.

In an endeavor to improve blood compatibility, the following surface modifications of polymers are currently being performed in an effort to mimic the functions of endothelial cells of blood vessels. This modification is limited to the blood-contacting surface region, while retaining mechanical integrity of the bulk polymers. The most practical approach is the immobilization and controlled release of bioactive agents such as heparin to prevent thrombus formation [8.4 – 8.7]. This technique has been applied to blood-contacting devices, and several heparinized catheters are now commercially available. Another approach is the bioinert surface, which could serve to reduce the interaction of blood components. One example is the incorporation of hydrophilic grafts such as polyethylene glycol onto hydrophobic polymer surfaces to minimize the protein adsorption [8.8, 8.9]. Yet another approach involves the concept of microphase-separated surfaces of hydrophilic-hydrophobic graft and block copolymers to regulate protein adsorption and cellular adhesion[8.10, 8.11]. In these last two methods, no bioactive agent is used. Nevertheless, these polymer surfaces have proved to exhibit excellent antithrombogenicity, and the application of these materials to blood-contacting devices such as catheters and small-diameter vascular grafts are now under way.

An approach with these combined techniques would provide great momentum towards increased blood compatibility of polymeric biomaterials and devices.

8.1.2.3
Polymer Membranes and Adsorbents as Biomaterials for Blood Treatment Applications

Extracorporeal blood treatment using membranes and adsorbents have become widespread, utilizing artificial kidneys, plasmapheresis, lymphocytapheresis and oxygenators. The techniques of blood purification treatment are all designed to remove pathogenic substances from the blood or correct the blood composition by diffusion, filtration, separation, adsorption or gas exchange. The most common application is membrane dialysis by means of the artificial kidney.

Table 8.1 shows the classification of currently used extracorporeal blood purification therapy using membranes and adsorbents. Besides hemodialysis, various new therapeutic methods have become available. A variety of hollow fiber membranes has been developed including cellulosics and synthetic polymers that are both hydrophilic and hydrophobic in nature, and their devices have become commercially available. Today, the hollow fiber type membrane device is widely used because of its compactness, ease of handling, high dialytic efficiency and low cost. The currently available hollow fibers may be classified depending on their membrane structures – homogeneous, asymmetric and microporous. The homogeneous membrane is suitable for hemodialysis and membrane oxygenation applications, the asymmetric membrane for hemofiltration, hemodiafiltration and continuous hemofiltration, and the microporous membrane for plasma separation, plasma filtration and membrane oxygenation. Since the membranes are used in contact with blood, their blood compatibility as well as mass transfer characteristics are of primary importance.

Remarkable advances have been made, particularly in the field of artificial kidney application. Today, end-stage renal failure patients numbering about 600 000 worldwide are reportedly undergoing hemodialysis treatment. Approximately 20 million m^2 of hollow fiber membranes are being used each year for hemodialysis in Japan alone, and worldwide consumption is estimated to be about three to four times greater. Recent advances in this type of membrane's applications are plasmapheresis involving double filtration and cryofiltration, capable of removing high molecular weight proteins like immunoglobulin and immune complexes with minimal loss of plasma. The usefulness of these membrane devices for blood purification treatment will be described in more detail in Sect. 8.2.

The extracorporeal adsorption method was first introduced by Yatzidis in 1964 with the use of activated charcoal for blood purification treatment of uremic patients [8.12]. Since then, various adsorbents, including activated carbon beads, ion exchange resins, uncharged resins and, recently, affinity adsorbents or immunoadsorbents based on immunoreactions to remove antibodies or immune complexes, have been developed and are now in use. The adsorption method has the advantage of a more selective removal of toxins or other pathogenic substances from the blood, compared with the membrane method. Adsorbents of activated carbon beads [8.13, 8.14] and anion-exchange porous resin [8.15], both coated with blood-compatible polymers such as PHEMA, have been used successfully for treatment of acute drug intoxication and hyperbilirubinemia, respectively. The polymer coating or microencapsulation is necessary for the purpose of preventing the release of microparticles from charcoals or porous resins, as well as improving their blood compatibility.

The affinity adsorbents are composed of polymeric carrier beads (e.g., cellulose or PVA gel, agarose) containing covalently immobilized affinity ligands of biologic (e.g., DNA, antigen, antibody, protein A) and non-biologic species. The use of nonbiologic, biomimetic ligands is recommended for clinical applications because of their sterilizability and non-antigenicity. The affinity adsorption with non-biologic ligand is explained as being due to physicochemical interactions such as ionic or hydrophobic binding. Typical examples of ligands currently available are hydrophobic, aromatic amino acids such as phenylalanine and tryptophan for the treatment of autoimmune diseases [8.16], and dextran sulfate for LDL removal in lipid metabolic disorders [8.17]. Recently, a novel immunoadsorbent with synthetic oligopeptide of acetylcholine receptor as an affinity ligand has been developed for the treatment of myasthenia gravis [8.18, 8.19], and a device incorporating it is becoming commercially available. Affinity adsorption therapy has the advantage that the specific removal of targeted substances from plasma can be performed without using replacement fluids such as expensive albumin preparations, which are required in membrane plasmapheresis. The clinical significance of this therapy has increasingly become recognized.

Finally, the use of polymeric materials in cell separation and cytapheresis is briefly described. Leucocyte removal is of great importance in therapy as well as blood tranfusion practice. Based on the adhesion property of leucocyte to fibrous materials, various fiber filters for leucocyte removal have been developed. Extracorporeal filtration lymphocytapheresis has been performed as an alternative to plasmapheresis by using a newly devised filter with non-woven fabrics of ultrafine polyester fibers (diameter 1.7 μm) for the treatment of autoimmune diseases and leukemia [8.20]. With this method, a more efficient leucocyte removal has been attained compared with the centrifugal method. Furthermore, the special hydrophilic polymer coatings of the fibers has been shown to offer more selective leucocyte removal with minimal loss of platelets [8.20, 8.21].

In order to develop polymeric adsorbents with specific affinity toward a particular subpopulation of lymphocytes, the molecular design of cellular specific polymers and their application to selective cell separation have been extensively studied [8.22–8.24]. A comb-type graft copolymer of PHEMA / polyamine has specific adsorption affinity toward B lymphocytes out of a mixture of B and T lymphocytes, and has been shown to provide excellent performance as a column adsorbent for the separation of B and T lymphocytes [8.22, 8.23]. The observed lymphocyte adsorption is explained as being due to the ionic interaction of protonated amino groups on the copolymer with acidic groups on the cellular plasma membrane. Furthermore, the partially guarternized copolymer has been shown to have a capability for separation of T cell subsets [8.24]. The application of such a copolymer to lymphocytapheresis will be of considerable interest.

Affinity separation of proteins and cells from blood is a promising frontier in the biomaterials field.

8.1.2.4
Polymeric Biomaterials for Ophthalmic Applications

Among biomaterials in ophthalmic applications, a contact lens (CL) is most widely accepted with outstanding impact on the general public. CLs currently available are divided into two types – hard CL(HCL) and soft CL(SCL). The basic requirements for CL materials are optical transparency, dimensional and thermal stability, suitable mechanical properties, oxygen permeability, surface wettability and biocompatibility.

In the history of CL development, polymer chemists have made significant contributions in developing sophisticated materials for CL application. In 1960, Wichterle introduced an innovative material of an acrylic hydrogel specifically designed for medical use [8.1], and, with slightly crosslinked PHEMA, made the first SCL. The HEMA-based SCL thus developed is hydrated to 38 % and has the advantage of comfort and permeability to water and oxygen, contrary to the traditional PMMA HCL. Since then, the PHEMA lens has become widely accepted and is commonly used all over the world. However, its oxygen supply to the cornea is insufficient when used as a continuous wear lens.

Over the past two decades, a number of SCL as well as HCL materials have been developed. The main aim in the CL materials research has been to achieve the high oxygen permeability required for continuous wear. Based on the direct correlation between water content and oxygen permeability, hydrogels with higher water content (> 70 %) have been extensively studied as an extended wear SCL material. These hydrogels have been prepared by using highly hydrophilic monomers such as N-vinylpyrrolidone (N-VP) or acrylamide, and ionic monomers such as methacrylic acid. At present, three types of hydrophilic SCL are commercially available, i.e., high water content (> 70 %) SCL, moderate water content (50–70 %) SCL, and low water content (10–50 %) SCL. The high water content SCLs have sufficient oxygen permeability (Dk value 60–70 × 10^{-11} (cm^2/sec) (ml O_2/ml · mm Hg) to allow wearing for extended periods of time. However, most of these SCLs have the disadvantage of poor mechanical properties and protein deposit build-up. As new candidates, collagen [8.25] and PVA hydrogel [8.26], now under development seem promising, because these hydrogels have excellent mechanical properties despite their extremely high water content (~90 %).

In addition to hydrophilic SCLs, hydrophobic SCLs made of elastomers such as silicone or butyl methacrylate-butyl acrylate copolymer are also available. They possess good oxygen permeability but poor surface wettability.

On the other hand, highly gas-permeable HCL materials have been developed by using copolymers of siloxanyl methacrylate with MMA, or copolymers of siloxanyl methacrylate with fluoro MMA, etc. [8.27]. These materials have been designed on the concept of hydridization of PMMA and silicone by utilizing the advantages of both materials. Recently gas-permeable HCL with highly increased oxygen permeability (Dk value 150–200) have become available.

In recent years, several disposable SCLs, based mainly on HEMA, have become commercially available. These disposable lenses, worn for a week or only a day and then discarded, reduce the problems associated with lens handling and care, and are recognized as better for eye health than conventional reusable extended wear SCLs.

The intraocular lens (IOL) based on PMMA has been used for four decades since the first implantation by Ridley of England in 1951 [8.28]. Today, cataract extraction with IOL implantation is considered one of the most successful surgical procedures. In the United States, IOL implantation is very popular, amounting to about 98 % of cataract operations, and currently over one million IOLs are reportedly implanted each year.

The IOLs widely used are based on PMMA or crosslinked PMMA with haptics, supporting parts of intra ocular lens; e.g. loops or clips of PMMA, polypropylene, 6-nylon or polyvinylidene fluoride (PVDF). The soft, foldable type of IOL from silicone, PHEMA, or fluoro-elastomer have been developed for small-incision surgery. In addition, multifocal IOLs have also become available.

The biostability and biocompatibility of implanted IOL materials have been intensively studied [8.29]. PMMA and PVDF have shown long-term biostability, while nylon 6 gradually underwent biodegradation. In an attempt to improve the long-term biocompatibility, surface modification of PMMA IOLs by heparin immobilization [8.30, 8.31] and glow-discharge treatment [8.32] have been performed. In vivo studies have shown improved biocompatibility as revealed by the significant reduction of cell-deposits on the modified surface [8.30–8.32].

With the rapid increase in the elderly population, demand for IOLs would grow, and thus the development of an ideal biocompatible IOL is very desirable.

As for the artificial cornea, there has also been great demand for it because of the shortage of cornea transplants. Attempts have been made to develop artificial corneas from biomaterials such as PMMA, polysulfone and PVA, but until now most of the clinical trials have failed because of rejection. Further study to explore more biocompatible materials is needed in the future.

8.1.3
Impact on Society

8.1.3.1
New Medical Technology and its Impact

Successful development of various medical devices in which macromolecules play a key role, as well as a large number of polymeric biomaterials, has previously been mentioned. Those medical devices have been developed with close cooperation between medical doctors, polymer chemists and engineers by using an interdisciplinary research team, and have been applied clinically after precise and lengthy governmental regulatory procedures (for example, Ministry of Health and Welfare in Japan, or Food and Drug Administration, in the United States). New therapeutic methods have been created, using those devices clinically accepted with high levels of success rate. These new medical technologies have had a tremendous impact on medicine because of their sometimes unique and unexpected therapeutic efficiency involving improved and sometimes new concepts. This impact obviously extends to patients and, in other words, to society. The nature of these impacts ranges from maintaining life of patients to esthetics of device wearers (patients). Impact may be classified as follows with corresponding devices.

a) Maintaining life of patient whose organ function is lost. Artificial kidney, intraaortic balloon pumping, artificial heart, ECMO, direct hemoperfusion.
b) Saving life of patient.
   Direct hemoperfusion.
c) Reducing serious symptoms of patient.
   Plasmapheresis, immunoadsorption, direct hemoperfusion, ascites treatment, intraocular lens.
d) Reducing invasiveness to patient body.
   Folded intraocular lens, dental adhesives.
e) Improvement in safety and side effects.
   Double filtration plasmapheresis, immunoadsorption, DDS, gas-permeable contact lens.
f) User (Patient) friendly.
   Contact lens, intraocular lens.
g) Reducing pain.
   Dental adhesives, transdermal delivery system.
h) Improvement of esthetics.
   Dental adhesives, contact lens.

Table 8.2 shows the new therapeutic methods and role of macromolecules in regard to those devices shown in the above classification.

Table 8.2 New therapeutic methods and role of macromolecules used in the new medical devices

Medical device	Therapeutic method	Role of macromolecule (requirement)
Artificial kidney	Extracorporeal hemodialysis (blood purification)	Membrane (permeation characteristics, and blood compatibility)
Intraaortic balloon pumping	Temporary cardiac assist (catheterization and pumping)	Balloon material (mechanical property and blood compatibility)
Artificial heart	Temporary and total heart replacement (implant)	Implant material (mechanical property and blood compatibility)
Oxygenator (ECMO)	Extracorporeal membrane oxygenation (blood purification)	Membrane (oxygen permeation and blood compatibility)
Direct hemoperfusion	Extracorporeal hemoperfusion (blood purification)	Coating material of activated charcoal or resin adsorbent (blood compatibility, permeation characteristics)
Plasmapheresis	Extracorporeal apheresis (blood purification)	Membrane (permeation characteristics and blood compatibility)
Plasma immunoadsorption	Extracorporeal plasma perfusion (blood purification)	Carrier in immunoadsorbent (blood compatibility)
Ascites treatment	Ascites collection, filtration, concentration and reinfusion	Membrane (permeation characteristics)
Intraocular lens	Replacement of lens after cataract operation	Lens and haptics (optical characteristics, and biocompatibility)
Dental adhesives	Filling of composite resin, metal bridging, crown fixation, core building up and lamination	Adhesive chemical (adhesive property with dentin, metal and resin)
Contact lens	Vision correction	Lens (optical characteristics, and oxygen permeability, or water content)
Drug delivery system	Controlled release and local delivery	Matrix or carrier (permeation characteristics, acceptability as drug and biocompatibility)

8.1.3.2
Establishment of Safety Standards

A wide range of clinical successes for new medical therapeutic methods using novel medical devices have opened up new business opportunities to industry. There have been entries into the medical device business by the big world corporations in the chemical and fiber fields, who have sufficient knowledge and know-how of the polymer industry. For example, Du Pont, Dow Chemical, Dow Corning, ICI, Bayer, AKZO and Rhône Poulenc have become involved in biomaterials or the medical device business using those polymeric biomaterials, or have continued in those businesses. In Japan, six fiber companies, Asahi Chemical, Toray, Teijin, Kuraray, Mitsubishi Rayon, and Toyobo have engaged in manufacturing hollow fiber membranes and in the hemodialysis business using those membranes. These entries have become a tremendous driving force towards further development of new medical devices.

However, new medical devices, which have shown clinically higher efficiency than ever before, have presented new safety problems either from using new biomaterials or by their exposure under new conditions of contact with human body tissue. Generally speaking, the safety of medical devices should be secured at both functionality such as performance or mechanical property, and biological toxicity. Their safety has had to be investigated from entirely new aspects because they have not previously existed in the medical fields of diagnosis or treatment. Ministry of Health and Welfare Japan requires the following four practices in order to prove the safety of new medical devices.

1) Evaluation of their safety before commercialization.
2) Management of their quality control on commercial production.
3) Monitoring their adverse effect after commercialization.
4) Re-evaluation of their safety.

Clinical and non-clinical study of new medical devices determines efficiency and safety of devices before commercialization. Biological toxicity evaluation is an essential part of non-clinical study. In order to establish common world standards for biological toxicity evaluation of medical devices, the International Organization for Standardization (ISO) has recently prepared the draft of TC194 entitled "Biological evaluation of medical and dental materials and devices" following several meetings of investigators from each country in the world, including U.S., Japan and EC countries since 1989. This will certainly affect regulatory biological toxicity evaluation standards for medical devices for most countries committed to TC194 of ISO.

Macromolecules used as biomaterials for these medical devices are usually prepared in the presence of monomer, polymerization catalyst, solvent, stabilizer, or some other additive. Residual amounts of these chemicals in the devices or chemical substances formed during polymerization or generated by decomposition during use in contact with human body tissue, or by exposure under sterilization conditions, often lead to biological toxicity. Biological toxicity testing includes cytotoxicity, mutagenicity, carcinogenicity, reproductive toxicity, irritation effects, sensitization effects, haemocompatibility effects, local effects of implantation and so on.

8.2
Application of Hollow Fiber Membrane Made from Various Macromolecules to Blood Purification Devices

8.2.1
Artificial Kidney

The artificial kidney is designed to replace kidney function by removing excess amounts of water and uremic toxins from the blood of the patient with renal failure.

Since the first dialysis experiment on dogs using collodion tube by Abel **et al** in 1913, and then the first successful clinical trial on a human patient with acute renal failure using cellophane tubing by Kolff in 1943, there have been many remarkable advances in the development of artificial kidneys, and they have had a great impact on medicine. In the early stages of dialyzer development, flat-plate and coil type modules using flat-sheet and tubular membranes, respectively, were used. However, with advances in membrane technology, hollow fiber type devices have become predominant. Today this type accounts for more than 95 % of all artificial kidneys.

Methods currently being used are hemodialysis (HD), hemofiltration (HF), hemodiafiltration (HDF) and continuous ambulatory peritoneal dialysis (CAPD). Hemodialysis offers a highly efficient removal of small molecules such as urea and creatine, and has been widely accepted for the treatment of renal failure patients. Hemofiltration shows excellent clearance of middle molecules, but low efficiency in removing small molecules, and requires a large quantity of infusion fluid. Hemodiafiltration removes both small and middle molecules efficiently, and the application of hemodiafiltration to short-time dialysis is being studied.

Hollow fiber membranes are roughly divisible into two categories: cellulosics and synthetic polymers. The regenerated cellulose membranes such as cuproammonium cellulose have been widely used up to now, mainly because of their high dialysis efficiency for small molecules, high mechanical strength in the wet state, and low production cost. On the other hand, in order to meet the need to improve clearance of middle molecules and blood compatibility, a variety of synthetic polymer membranes such as EVAL, PAN, PMMA, PS, PA and PEPA, as shown in Table 8.1, have been developed and are gaining wider acceptance. In general, hemodialysis uses hydrophilic homogeneous dense membranes, and hemofiltration and hemodiafiltration use hydrophobic asymmetric membranes.

With the increase in the population of long-term (over 5–10 years) dialysis patients, complications including anemia, bone or joint disorder, and peripheral neuropathy have arisen to a significant extent. The cause of these complications has been thought to be due to the gradual accumulation of middle and large molecules not removed by the conventional membrane such as cellulose. In fact, it has been demonstrated that some dialysis complications, such as anemia and bone pain, have been successfully treated with hemodialysis using an EVAL membrane with a pore size large enough to allow protein permeation to an appreciable extent (protein permeating hemodialysis) [8.33]. Recently, it has been reported that the accumulation of β_2-microglobulin (mol. wt. 11 800) is the cause of amyloidosis observed for many long-term dialysis patients

[8.34]. Besides the EVAL membrane, high-performance membranes with high permeability to low molecular weight proteins (mol. wt. 10 000–30 000), in particular β_2-microglobulin, have been intensively developed using synthetic polymers such as PMMA, PS and PAN, and are now in clinical use.

The safety of membranes and their blood compatibility are of great importance. When blood comes in contact with a membrane, various reactions take place in the coagulation and immune systems. During the past decade, there has been an increasing awareness of the importance of the interactions of blood and membranes. Therefore, the immune reactions, including the complement activation and antithrombogenicity of membranes, have been studied extensively.

Transient leucopenia, a phenomenon of rapid decrease of white blood cell level, is observed about 15–30 min after the start of dialysis [8.35]. It has been shown that this is caused by complement activation arising from contact between blood and membrane [8.36]. The degree of complement activation varies with the membrane material. In general, the cellulosic membrane activates the complement system more strongly than synthetic polymer membranes. Many arguments have been made about the relation of complement activation to immunodeficiency in dialysis patients, and with acute symptoms such as allergic reaction, anaphylactic shock, itching, and hypotension during dialysis.

In ordinary hemodialysis treatments, an anticoagulant such as heparin is used to prevent blood clotting. However, heparin is not allowed to be used in immediate post-operative patients and those with bleeding risk factors, and long-term use of heparin is said to develop some side effects, such as abnormalities of lipid and bone metabolism and allergic reaction. Of the various membranes, the EVAL membrane has good antithrombogenic properties, allowing dialysis treatment without using any anticoagulant, and is currently being used for treating a variety of dialysis patients, including those with a high bleeding risk factor [8.37]. The antithrombogenicity of the EVAL membrane is thought to be due to its unique molecular structure arising from the copolymer of hydrophilic vinyl alcohol and hydrophobic ethylene, and the selective adsorption of albumin on the membrane surface [8.38].

Efforts have been made to improve blood compatibility of membranes. For instance, with cellulose membranes, the surface modification has been done in such a way that the hydroxyl groups on the surface, considered to be responsible for complement activation, are partially substituted by diethyl aminoethyl groups [8.39] or covered with cationic polymers containing amino groups [8.40]. Moreover, the grafting of polyethylene glycol [8.9] or phospholid polymer [8.41] onto cellulose membranes has shown to provide good antithrombogenicity as well as significant suppression of complement activation. Some of these modified membranes are now commercially available.

Recently, the release of interleukin-1, tissue necrosis factor and platelet activating factor, and superoxide generation have been shown to be involved in the blood/membrane interactions, and their relations with the observed complications in dialysis patients are being studied [8.42].

As a new development, continuous arterio-venous hemofiltration (CAVH) has been developed and used for the treatment of acute renal failure [8.43]. This method uses a hemofilter with a small membrane surface area to perform blood filtration continuously for several days, utilizing the arterio-venous pressure difference. The CAVH method is similar to glomerular filtration and is expected to be applied to a wearable type of artificial kidney using an antithrombogenic hollow fiber membrane in the future.

Advances in treatment by artificial kidneys has enabled many renal failure patients to live for over twenty years, offering a success record comparable to kidney transplantation. To improve the quality of life of long-term dialysis patients, continuous efforts have to be devoted to the developments of more biocompatible membranes and devices.

8.2.2
Membrane Plasmapheresis

Plasmapheresis is a blood processing method for all plasma components, including large molecules with a molecular weight of a few million Dalton, and is used for the treatment of various intractable diseases. The substances to be removed by plasmapheresis include: auto-antibodies and immune complexes in autoimmune diseases such as rheumatoid arthritis, systemic lupus erythematosus, or myasthenia gravis; immunoglobulin M in hematological diseases such as macroglobulinemia; protein-bound substances in drug poisoning or fulminant hepatitis; and cholesterol in hyperlipidemia.

There are two types of plasmapheresis procedures available: plasma exchange which separates the patient's blood into blood cell and plasma components using a plasma separator, replacing the plasma component with healthy human plasma; and plasma processing which selectively removes pathogenic substances from the separated plasma using a membrane or adsorbent, as already shown in Table 8.1.

The plasma separator consists of microporous membranes, with a pore size of 0.2 to 0.5 µm to separate all blood cellular components ($> 2 - 3$ µm) and plasma components. At present, there are many kinds of hollow fiber membranes available in the market for plasma separation: hydrophilic PVA, hydrophobic cellulose acetate, PP, PE, PMMA, and PS (see Table 8.1).

Plasma exchange is a simple and efficient technique, but it requires a large quantity of expensive replacement fluid (fresh frozen plasma or albumin preparations) and there is also a possible risk of infection, such as HIV or hepatitis virus from the replacement fluid. To solve these problems, double filtration or cascade filtration has been developed based on the principle of sequential filtration, using two filters with different pore sizes: a first filter (plasma separator) and a second filter (plasma fractionator) for selective removal of large molecules, including pathogenic substances, from the plasma [8.44].

The plasma fractionation membrane has a pore size from 0.01 through 0.05 µm, and EVAL, CA or PE hollow fiber membrane is used. With EVAL membranes, four types with different pore sizes 2A, 3A, 4A and 5A are available. The 4A and 5A membranes are suitable for removing IgM (mol. wt. 950 000) fraction and β-lipoprotein (mol. wt. 2 000 000). These membranes showed good sieving of albumin (mol. wt. 67 000) and removed sufficient harmful substances as well [8.45]. To separate IgG (mol. wt. 160 000) fraction from albumin, the 2A membrane seems to be suitable.

As a different approach to the plasma fractionation, cryofiltration [8.46] has been developed for selective removal of solutes by cooling withdrawn plasma at 4 °C to form cryogel, which is then removed by secondary filtration.

The clinical efficacy of these membranes' plasmapheresis has been demonstrated in the treatment of the aforementioned intractable diseases.

In extensive studies on blood/membrane material interactions for improving blood compatibility of membranes, much attention has recently been paid to the influence of membrane materials upon immune response, and the importance of immunomodulation has been recognized [8.47]. It is expected that a specific interaction between blood and membrane may serve to play a beneficial role in the extracorporeal treatment of immunological disorders such as autoimmune diseases. Further study is now in progress.

8.2.3
Membrane Oxygenator

The oxygenator is an artificial lung device which oxygenates venous blood extracorporeally during open-heart surgery. It is roughly divisible into two types: bubble and membrane types. The bubble oxygenator performs the exchange of gases by direct contact of blood with oxygen, while, in the membrane type, the gas exchange takes place through the gas-permeable membrane. Historically, the bubble oxygenator has been preferred for a long period of time, although it is now being displaced by the membrane type mainly because of problems of erythrocyte destruction and protein denaturation. Today, a number of gas-permeable hollow fiber membranes have become available. At present, the total number of artificial lungs used worldwide each year is reportedly about 540 000. The proportion of membrane oxygenators used has rapidly increased, accounting for almost 95 % especially in the United States. The membranes fall basically into the following two categories: homogeneous membrane, for example, silicone membrane, which allows gas transfer by solubility and diffusion; and microporous polyolefin membrane such as PP membrane, which allows gas transfer through its micropores. The silicone material is highly gas-permeable, but has poor mechanical strength. On the other hand, the microporous PP membrane offers a high gas permeability when used for ordinary open-heart surgery (2–3 h), but, if used for a long period of time, causes serum leakage due to protein deposition on the surface, with a resultant rapid reduction in gas exchange efficiency. To prevent this problem, its micropores are filled with silicone or fluoropolymer, or the whole membrane surface is thinly coated with such polymers to form a composite membrane. Microporous membranes thus treated are free from serum leakage and offer a better gas permeability compared to the ordinary silicone membrane.

Recently, a hollow fiber membrane with an asymmetric microporous structure of new polyolefin has been successfully manufactured by a special melt spinning process and applied to an oxygenator [8.48]. The membrane has a skin layer on its outer surface, which is suitable for external perfusion of blood. The skin layer is so thin, less than 1 μm, that it is highly permeable to gases and causes no serum leakage if used for a long-term oxygenation. With use of this unique membrane, a high performance, ultra-compact oxygenator has been developed and is now in clinical use.

For the purpose of a long-term assistance in the treatment of respiratory failure patients, extracorporeal membrane oxygenation (ECMO) has been developed and is now in clinical use. Since the ECMO device is used for long term treatment, the membrane is required to cause no serum leakage and to have good blood compatibility. Therefore, the new polyolefin membrane with asymmetric structure [8.49] and PP-silicone composite membrane are suitable.

In addition, a new device, called an intravascular oxygenator (IVOX), has recently been developed and is in clinical trials [8.50]. It is interesting that IVOX utilizes the concept of intravenous oxygen delivery with the use of microporous PP hollow fibers, mimicking the natural exchange of gases in the blood.

8.2.4
Other Devices

In addition to the above-mentioned applications, hollow fiber membranes have also been used in such fields as ascites treatment, hemoconcentration for autologous blood recovery, donor plasmapheresis and removal of AIDS and hepatitis viruses from blood.

Most of the blood treatment membranes currently in use are made of general purpose polymers originally developed for industrial use. There have been no truly blood-compatible membranes. Therefore, it is desirable to develop a tailor-made, biocompatible membrane with high performance and selectivity so as to meet a specific need of intended applications.

8.3
Impact of Dental Polymeric Adhesives on Dental Materials and Dentistry

8.3.1
Dental Polymeric Adhesives

Technology in dentistry has been significantly dependent on the advances of material science and engineering. Since a tooth cannot regenerate, the use of artificial materials such as metals, ceramics and polymeric materials is needed to fix tooth defects. With recent advances in polymer chemistry, many polymeric materials have come into wide use in various applications as prosthetic devices like denture base, plastic teeth, and as composite restorative filling materials, bonding agents, cements, pit and fissure sealants, and impression materials.

As a restorative filling material, amalgam has long been used. However, in recent years, polymeric resin materials have been increasingly used, mainly because their high transparency make it possible to produce esthetic restorations that are compatible with natural tooth.

The first dental resin was an acrylic resin, developed in the 1940s. This resin was produced by mixing PMMA powder with MMA monomer, but had some clinical problems such as poor strength and low abrasion resistance. In the early 1960s, another type of resin, called composite resin, was developed. This new type of resin contained a high percentage of inorganic filler in order to improve its physical properties such as strength, hardness and resistance to abrasion. However, the resins available at that time had no adhesive properties with tooth structure, and they therefore required the placement of undercuts and retentive grooves, at the sacrifice of healthy tooth structure, for mechanical retention.

In addition to physical properties, adhesion to tooth structure is of prime importance in clinical applications. When a filling material bonds poorly to tooth structure, it would not only fall from the cavity in a short time, but also produce a gap between the cavity wall and filling material, thus causing microleakage. This will eventually lead to severe defects such as pulpal irritation, secondary caries and marginal discoloration due to the penetration of bacteria and corrosive materials. Although the adhesion of filling material to tooth structure was strongly desired, sufficient bond strength to tooth structure has not been available until recently because of the following special conditions in the oral environment.

(1) The oral cavity is always wet due to saliva, so that it is difficult to bond the resin to tooth structure.
(2) Oral temperature varies rapidly within the range 0–50 °C when food and drink are taken in. This produces a difference in thermal coefficient of expansion between the filling material and tooth structure, often leading to the destruction of adhesion.
(3) The resin tends to get deformed due to concentrated stress because it is constantly subjected to occlusal stress as high as 20–100 kg/cm^2, thus often leading to the destruction of adhesion.

Adhesion of filling materials to dentin and enamel has really been a major research objective for many investigators for over a quarter of a century. In fact, a number of studies have been made in pursuit of adhesive restorative materials which are capable of producing biocompatible interface between material and tooth structure.

There are three main approaches to bonding the resin to tooth structure.

1) Mechanical interlocking.
2) The use of functional monomers which show high affinity to inorganic substances (hydroxyapatite) and/or organic substances (collagen).
3) The use of wetting agents which improve the compatibility of resin with tooth structure.

One method that falls in the first category is the acid etching technique by Buonocore [8.51], by which the tooth enamel is acid etched with phosphoric acid to roughen its surface. As an etchant, nitric acid, citric acid or oxalic acid is also used. The acid-etched surface provides a microstructure of enamel rods, which is capable of mechanical interlocking. The functional monomers mentioned in category 2) have often been used as adhesion promoting monomers to improve the adhesion between tooth structure and the resin. Figure 8.2 shows typical examples of monomers promoting adhesion to tooth structure. These monomers are characterized by their functional

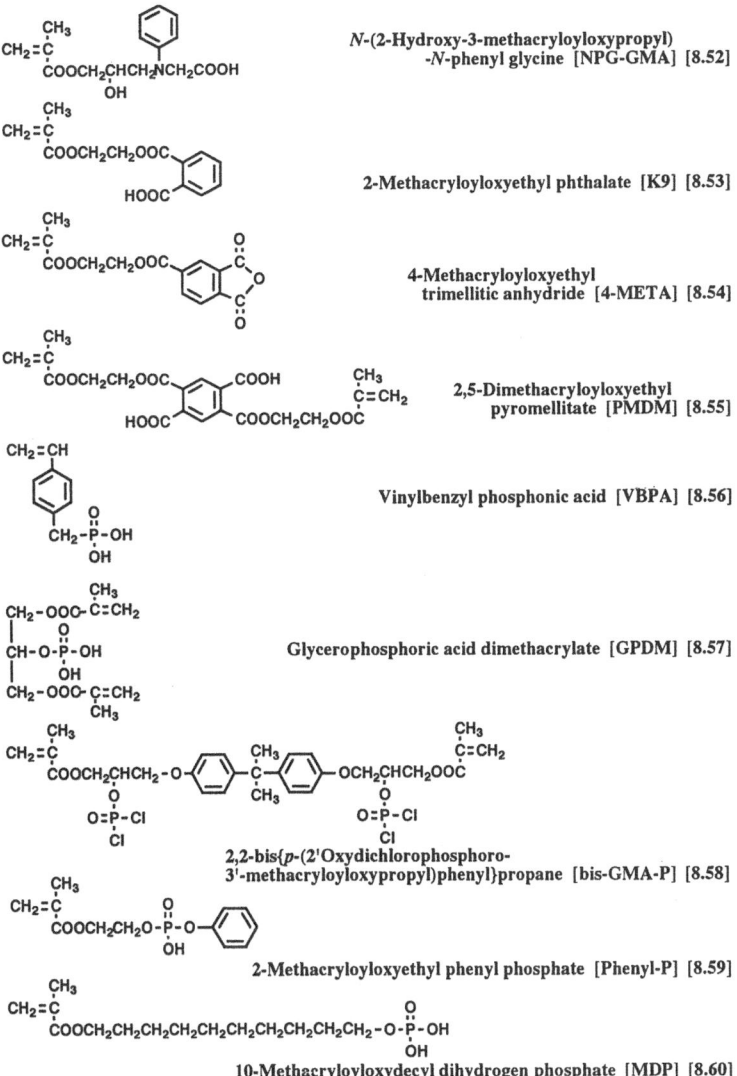

Fig 8.2 Adhesion-promoting monomers to tooth structure

groups, such as phosphate group, or carboxylate group, which are believed to interact with tooth structure by hydrogen bonding or chelate formation. As a wetting agent in category 3), hydrophilic monomers such as HEMA are preferable.

Historically, Sevriton (Amalgamated Dental, England) was the world's first adhesive resin, developed in 1952 by using GPDM [8.57] as an adhesion promoting monomer. Sevriton attracted considerable attention as the first dental adhesives among researchers, but its adhesiveness was so low and unstable that it did not satisfy any practical use. In 1977, Clearfil Bond System F (Kuraray, Japan) was developed, which used Phenyl-P [8.59] as an adhesion promoting monomer. Phenyl-P acts as a bonding

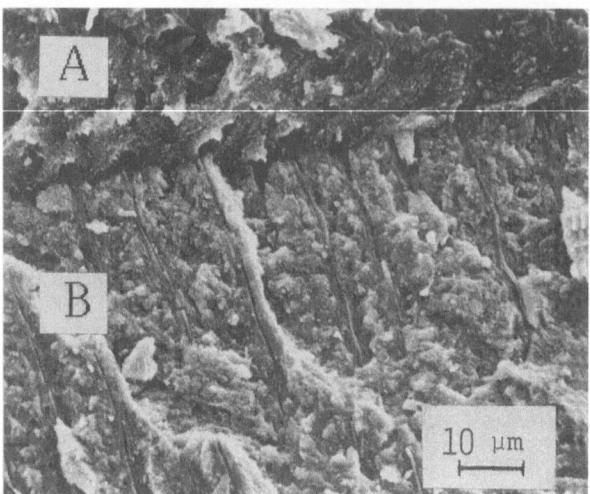

Fig 8.3 SEM picture of adhesive bond interface in cross section
A: cured composite resin (Clearfil Bond System F)
B: dentin

agent to promote adhesion between Clearfil composite resin and tooth substrates. This bonding system gained wide acceptance with many dentists because of its excellent adhesion to tooth structure as well as its clinical practicability. Figure 8.3 shows SEM picture of an adhesive bond interface, in cross section, of a joined specimen obtained by the use of Clearfil Bond System F. At the interface between the two phases of cured resin (A) and dentin (B), thin layer of adhesive resin is seen, indicating that resin penetrated deeply into dentinal tubules and bonded tightly to dentin tissue. The high bond strength can be explained as due to not only resin tag formation into dentinal tubules, but also to so-called resin-reinforced dentin layer formed at the interface. Typical dental polymeric adhesives developed during the past four decades are shown in Table 8.3. While dental adhesives systems of the early period offered a bond strength as low as $20 \, \text{kg}/\text{cm}^2$, some recent adhesives, such as Clearfil New Bond and Orthomite Super Bond, provide markedly improved bond strength as high as $100 \, \text{kg}/\text{cm}^2$ or more.

Dental filling materials are generally composed of a low viscosity primer or bonding agent which includes adhesive monomer to promote adhesion to dentin and enamel, and a high-viscosity composite resin to provide the physical properties. Composite resin contains high concentrations of inorganic filler dispersed in a resin matrix of bifunctional methacrylate monomers like Bis-GMA, the adduct of bisphenol A and glycidyl methacrylate in most cases, in order to attain optimal physical properties and to minimize polymerization shrinkage. The content of the filler depends upon its particle size and distribution, but is usually 75–85 % by weight. The filler particles are usually treated with silane coupling agents to enhance their compatibility with the resin matrix. As a polymerization initiator, a redox system using benzoyl peroxide/amine is usually employed because the resin needs to cure in a few minutes

Table 8.3 Typical dental polymeric adhesives products

Product	Manufacturer	(Country)	Adhesion promoting monomer
Sevriton	Amalgamated Dental	(England)	GPDM
Cosmic	Amalgamated Dental	(England)	NPG-GMA
Clearfil Bond System F	Kuraray	(Japan)	Phenyl-P
Orthomite Super Bond	Sun Medical	(Japan)	4-META
Tenure	Den-Mat	(USA)	NPG-GMA/PMDM
Scotchbond	3M	(USA)	Bis-GMA-P
Clearfil New Bond	Kuraray	(Japan)	MDP

at room temperature. However, when acidic monomers are used as adhesive components, the curing of the resin is retarded because the amine is neutralized. It is thus important to find a suitable initiator which does not inhibit the curing of the resin with acidic monomers. Clearfil Bond System applies benzoyl peroxide / amine / salt of sulfinic acid [8.61] as a polymerization initiator which plays a key role in improving adhesion. Recently a photo-curing system has been introduced instead of the conventional redox system, and has now become widely accepted because of its easy handling and high performance in various restorations.

8.3.2
New Dental Materials and New Dentistry Based on Dental Adhesives

With recent advances in dental adhesive technology, a variety of dental restorative techniques have been successfully developed in clinical applications over and above adhesive restorative filling. For instance, the restoration of metal crowns have long been made with core build-up using metal, but this method has some shortcomings, such as a long casting time of the metal core or insufficient core retention due to poor bonding between tooth structure and metal core. Recently, however a new restoration technique, called the composite core build-up system, has been developed, in which a metal post is implanted in the dental root and the core is built up with adhesive composite resin. This new technique is rapidly gaining full acceptance because of its easy handling and delivery of high retention force.

In addition, there has been considerable demand for high bond strength between metal and tooth structure in many dental restorations such as metal crowns and metal inlays. In order to produce a strong bond between the resin and metal, many approaches have been made using adhesive monomers which are considered to provide a chemical bond to both metal and tooth structure. Among adhesion promoting monomers shown in Fig. 8.2, 4-META [8.54] and MDP [8.60] provide excellent adhesion to metal, and new adhesive cements based on these monomers have been developed. With the advent of such new adhesive cements, the so-called adhesion bridge technique, which retains the bridge with bonding between tooth structure and metal while minimizing the cutting of healthy tooth structure, has been established. This adhesion bridge technique is increasingly accepted as an alternative to the conventional bridge construction system using mechanical interlocking, which requires cutting of any excess of tooth substrate.

On the other hand, composite resins which offer greater strength and higher abrasion resistance have become available with the development of new filler formulations and effective surface treatment technology as well as the improvement of the matrix resin. Recently, these composite resins are increasingly used for replacement of conventional porcelain and metal in dental restorations. In particular, the composite restoration technique has found much wider application with the introduction of a photo curing system, owing to its easy manipulation and effective performance. In new applications, composite resin has been used for veneered crown and laminate veneer restoration instead of porcelain. It has also been used for direct and indirect inlay restoration instead of metal. Such techniques allow the dentist to produce esthetic restorations, creating new dental therapeutic procedure called esthetic dentistry. In addition, the adhesive technology has been applied to pit and fissure sealants for prevention of caries. A new type of pit and fissure sealant using the adhesive monomer (MDP) and a newly developed fluoride-releasing polymer has become available [8.62]. This product offers superior adhesiveness and slow release of fluoride over a long period of time.

The new dental adhesive materials developed during the last decade have been leading a conceptual and technical revolution in dental treatment, creating a new discipline called adhesive dentistry. With future advances in polymer chemistry, it is expected that more promising dental materials, such as resin systems, with zero polymerization shrinkage or slight expansion upon polymerization, will be developed, and contribute to adhesive dentistry in the years ahead.

8.4
References

8.1   Wichterle O, Lim D (1960) Nature 185: 117
8.2   Ikada Y (1991) Polymer J (Tokyo) 23: 551
8.3   Matsuda T (1987) Jpn J Artif Organs 16: 1252
8.4   Tanzawa H, Mori Y, Harumiya N, Miyama H, Hori M, Ohshima N, Idezuki Y (1973) Trans Amer Soc Artif Inter Organs 19: 188
8.5   Ito Y (1987) J Biomater Appl 2: 235
8.6   Jacobs HA, Okano T, Kim SW (1989) J Biomed Mater Res 23: 611
8.7   Nojiri C, Park KD, Grainger DW, Jacobs HA, Okano T, Koyanagi H, Kim SW (1990) Trans Amer Soc Artif Inter Organs 36: M168
8.8   Mori Y, Nagaoka S, Takiuchi H, Kikuchi T, Noguchi N, Tanzawa H, Noishiki Y (1982) Trans Amer Soc Artif Inter Organs 28: 459
8.9   Akizawa T, Kino K, Koshikawa S, Ikada Y, Kishida A, Yamashita M, Imamura K (1989) Trans Amer Soc Artif Inter Organs 35: 333
8.10 Nakashima T, Takakura K, Komoto Y (1977) J Biomed Mater Res 11: 787
8.11 Okano T, Nishiyama S, Shinohara I, Akaike T, Sakurai Y, Kataoka K, Tsuruta T (1981) J Biomed Mater Res 15: 393
8.12 Yatzidis H (1964) Proc Eur Dial Transplant Assoc 1: 83
8.13 Nakashima T, Takakura K, Komoto Y, Nakabayashi N, Inou T (1979) Jpn J Artif Organs 8: 460

8.14 Inou T, Otsubo O, Ota K, Agishi T, Nakagawa S, Nakabayashi N (1979) Jpn J Artif Organs 8: 505

8.15 Nakaji S, Inukai Y, Takakura K (1993) in: Agishi T, Kawamura A, Mineshima M (eds) Therapeutic Plasmapheresis vol 12, VSP, Utrecht Tokyo, p 211

8.16 Yamazaki Y, Yamawaki N (1983) Jpn J Artif Organs 12: 895

8.17 Yokoyama S, Hayashi R, Kikkawa T, Tani N, Takada S, Hatanaka K, Yamamoto A (1984) Atherosclerosis 4: 276

8.18 Tanihira M, Oka K, Nakashima T, Okumura S, Ide Y, Takamori M (1989) Jpn J Artif Organs 18: 15

8.19 Nakaji S, Oka K, Tanihara M, Takakura K, Takamori M (1993) in: Agishi T, Kawamura A, Mineshima M (eds) Therapeutic Plasmapheresis vol 12, VSP, Utrecht Tokyo, p 573

8.20 Umegae M, Nishimura T, Kuroda T, Kato H (1988) Jpn J Artif Organs 17: 413

8.21 Kinugasa E, Takayama K, Akizawa T, Koshikawa S, Mori Y, Kuroki Y, Nishimura T, Kuroda T (1989) Jpn J Artif Organs 18: 1346

8.22 Kataoka K, Sakurai Y, Tsuruta T (1987) in: Brash JL, Horbett TA (eds) Proteins at Interfaces: Physicochemical and biochemical studies, ACS Symp Series No 343, Am Chem Soc, Washington DC, p 603

8.23 Kataoka K (1988) Artificial Organs 12: 511

8.24 Kataoka K (1991) Jpn J Artif Organs 20: 314

8.25 US Patent 4,260,228 (April 7, 1981), Miyata T (to Opticol Corp)

8.26 Kita M, Ogura Y, Honda Y, Hyon S, Cha W, Ikada Y (1990) Graefe's Arch Clin Exp Ophthalmol 228: 533

8.27 Arkles B (1983) Chemtech Sept 542

8.28 Ridley H (1951) Trans Ophthalmol Soc UK 71: 617

8.29 Yamanaka A, Matsumoto T, Nakamae K, Kanaji Y, Goto H, Kitayama M (1979) Am IOL Soc J 5: 131

8.30 Larsson R, Selen G, Björklund H, Fagerholm P (1989) Biomaterials 10: 511

8.31 Spångberg M, Kihlström I, Björklund H, Bjurström S, Lydahl E, Larsson R (1990) J Cataract Refract Surg 16: 170

8.32 Ohira A, Oshima K, Yamanaka A, Goto H, Nakamae K (1986) Acta Soc Ophthalmol Jpn 90: 1591

8.33 Saito A, Naito H, Hirohata M (1983) Progress in Artif Organs 412

8.34 Gejyo H, Yamada T, Odani S, Nakagawa Y, Arakawa M, Kunitomo T, Kataoka H, Suzuki M, Hirasawa Y, Shirahama T, Cohen AS, Schmid K (1985) Biochem Biophys Res Commun 129: 701

8.35 Kaplow L, Goffinet J (1968) JAMA 203: 1135

8.36 Craddock P, Fehr J, Dalmasso A, Brigham K, Jacob H (1977) J Clin Invest 59: 879

8.37 Naito H (1981) Artificial Organs 5 (Suppl. 1): 670

8.38 Sakurada Y, Sueoka A, Kawahashi M (1987) Polymer J (Tokyo) 19: 501

8.39 Bosch T, Schmidt B, Samtleben W, Gurland HJ (1986) Clin Nephrol 26 (Suppl. 1): 22

8.40 Akizawa A, Kitaoka T, Koshikawa S, Watanabe T, Imamura K, Tsurumi T, Suma Y, Eiga S (1986) Trans Am Soc Artif Inter Organs 32: 76

8.41 Ishihara K, Fukumoto K, Aoki J, Nakabayashi N (1992) Biomaterials 13: 145

8.42 Akizawa T (1991) Jpn Society for Dialysis Therapy 24: 433

8.43 Krammer P, Böhler J, Kehr A, Gröne HJ, Schrader J, Matthaei D, Scheler F (1982) Trans Am Soc Artif Intern Organs 28: 28

8.44 Agishi T, Kaneko I, Hasuo Y, Hayasaka Y, Sanaka T, Ota K, Amemiya H, Sugino N, Abe M, Ono T, Kawai S, Yamane T (1980) Trans Am Soc Artif Intern Organs 26: 406

8.45 Sueoka A, Takakura K (1991) Polymer J (Tokyo) 23: 561

8.46 Malchesky PS, Asanuma Y, Smith JW, Kayashima K, Zawicki I, Werynski A, Blumenstein M, Nosé Y (1981) Trans Am Soc Artif Intern Organs 27: 439

8.47 Omokawa S, Malchesky PS, Goldcamp JB, Savon SR, Nosé Y (1991) J Biomed Materials Res 25: 621

8.48 Akasu H, Anazawa T (1990) J Jpn Soc Biomaterials 8: 141

8.49 Tatsumi E, Taenaka Y, Nakatani T, Akagi H, Sekii A, Yagura A, Sasaki E, Goto M, Nakamura A, Takano H (1990) Trans Am Soc Artif Intern Organs 36: M480

8.50 Mortensen J, Barry G (1989) Int J Artif Organs 12: 384

8.51 Buonocore MG (1955) J Dent Res 34: 849

8.52 Bowen RL (1965) J Dent Res 44: 895

8.53 Masuhara E, Kojima K, Tarumi N, Nakabayashi N, Hotta H (1967) Rep Inst Med Dent Eng Tokyo Med Dent Univ 1: 29

8.54 Takeyama M, Kashibuchi N, Nakabayashi N, Masuhara E (1978) J Jpn Soc Dent Appar Mater 19: 179

8.55 Bowen RL, Cobb EN, Raposon JE (1982) J Dent Res 61: 1070

8.56 Anbar M, Farley EP (1974) J Dent Res 53: 879

8.57 Buonocore MG (1956) J Dent Res 35: 846

8.58 US Patent 4,482,505 (November 13, 1984) , Bunker JE, Lake WB (to 3M Corp)

8.59 Yamauchi J, Nakabayashi N, Masuhara E (1979) ACS Polymer Preprints 20: 594

8.60 Omura I, Yamauchi J (1989) Trans. of International Congress on Dental Materials p 356

8.61 US Patent 4,182,035 (January 8, 1980), Yamauchi J, Yamada K, Shibatani K (to Kuraray Co. Ltd.)

8.62 Nishida K, Yamauchi J, Kadoma Y, Masuhara E (1985) Rep Inst Med Dent Eng Tokyo Med Dent Univ 19: 21

Agendas for Research Cooperation in the Japanese Chemical and Materials Industry

NAOYA YODA

9.1
Abstract

Chemistry plays a significant role in creating new materials because it manipulates and creates new substances at the molecular level. The twenty-first century is regarded as the new chemical age, and the new advanced materials are characterized as designated materials on the molecular level. Based on long-range research and development experience at Toray Industries, the following key points are important for the development of new advanced materials:

1) Strong emphasis on basic research with innovative concepts.
2) Direct risk-taking by corporate executives.
3) Open-minded joint work with users for developing applications.

The new advanced materials are expected to satisfy human needs in futures, and the chemical industry should respond to this growing expectation by research cooperation among competitors in a precompetitive stage.

The following topics are discussed:

1) Industrial trends in the Japanese chemical and materials industries.
2) Competition and cooperation (C & C). Public and private cooperation in R & D activities for innovative future technology.
3) Setting research agendas and organization of the public and private sectors in the Japanese national plan.
4) Perspectives for the chemical industry in the new chemical age.

It should be stressed that the national project was successfully introduced in Japan by the private initiative of an informal working relationship between the industrial and public sectors.

Secondly, it should be emphasized that governments, individually and collectively, can enhance such efforts by creating well-balanced competition and cooperation, a climate of sustained economic growth, by supporting basic research and education. Free competition in the R & D of innovative technology is invaluable. In private industry, short-range R & D to modify existing products in the market is emphasized, whereas the R & D of both the public and private sectors should be focused on technology and new materials for the long-range future. In authors's view, private initiatives not directly by the government are the way to bring about well-balanced

Okamura, Rånby, Ito (Eds.): Macromolecular Concept and Strategy
© Springer-Verlag Berlin · Heidelberg 1996

cooperation and competition in R & D for future technology. If this can be brought about, the author foresee a bright future, not just for the chemical industry, but for all sectors of industry. And, best of all, the chief beneficiary of that future will be international society at large. [9.2]

The key issues may be summarized as follows:

1) Excessive competition and duplication of R & D should be eliminated.
2) Unless a given industrial sector is prosperous, individual firms in the sector will not survive.

Based on my experience during 1976–1988 with the government, the Federation of Management of Chemical Industry, Japan (Kobunshi-Doyukai) and the Japanese Round Table (Keidanren), the author will explore the following topics:

1) Industrial trends in the Japanese chemical and materials industries.
2) Public and private cooperation in R & D activities dealing with innovative future technology.
3) Setting research agendas and the organization of research by the public and private sectors – the uniqueness of the Japanese system.
4) Perspectives for the chemical industry in new chemical age.

This chapter emphasizes the importance of private initiatives in the initial informal working relationship with public sectors in 1977–1979 to set up the process of formulating industrial policy by identifying targets and then setting research agendas. In early 1977, the private sector achieved consensus on the basic framework of the project and proposed a draft industrial policy to the government. The author will give readers his impression of U.S.-Japanese competition, based on ten years in the U.S. as a student, scholar, and business executive. In addition, the author will cite an example of an official project concerning the basic technology industries. [9.3]

9.2
Industrial Trends and Restructuring in the Japanese Chemical and Materials Industry

When the chemical industry is envisaged in the framework of Japan's the long-term future, it is urged to understand and put into practice the policies proposed by the Council of Economic Structural Adjustment for International Cooperation. The concepts of the two Maekawa Reports of May 1985 and the new Maekawa Report of May 1987 are shown in Fig. 9.1 below:

These reports describe how the Japanese structure will change in the period from 1985 to 2000. The manufacturing division is expected to shrink, while the service sector will keep expanding. Annual GNP growth is forecast to drop from the 3.69 % achieved in 1985 to 1–1.5 % in 2000.

The new Maekawa report forecasts that the percentage of manufacturing industry in the GNP will drop from 30.1 % in 1985 to 26.7 % in 2000. Above all, the materials industries, including steel, nonferrous metals, and cements, as well as the chemical industry, are expected to experience severe recession. The system assembly industry

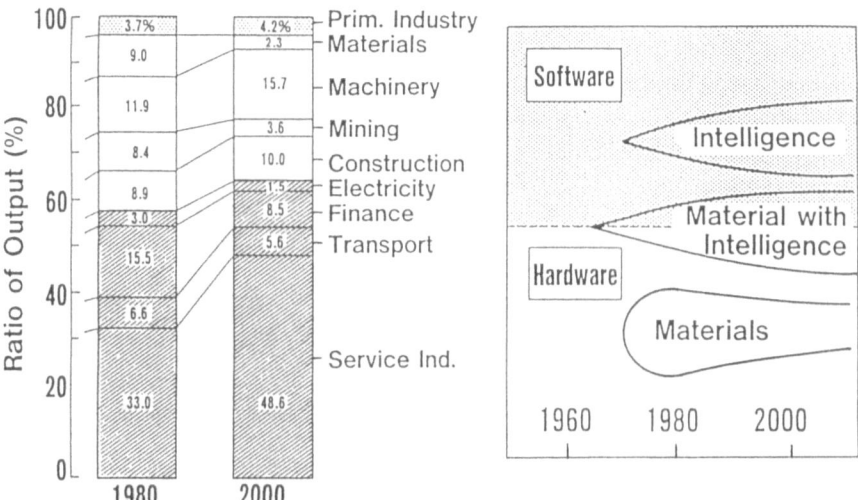

Fig. 9.1 Structural change of Japanese industries

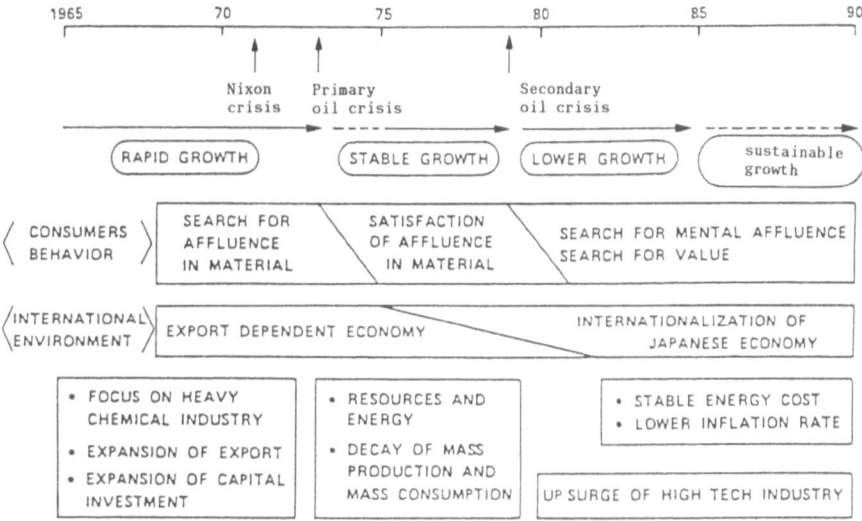

Fig. 9.2 Trends of the Japanese economy

and electronics will continue high growth, but automotive-related business will be getting smaller. The local production ratio of the manufacturing industry in overseas markets should increase from 4.3 % in 1984 to 8.2 % in 1993. However, exports ratio of manufacturing industry from Japan will decline from 13.5 % to 11.8 %, while the import ratio of domestic production would increase from 5.4 % to 6.8 %. By 2000, the economy is expected to grow 4 % annually; during the next fifteen years, the GNP would increase 1.5 times. The surplus current balance to the GNP would be dropping from 3.6 % in 1985 to 2 % in 1993 and 1–1.5 % in 2000. The effective exchange rate is assumed to be about $ 1 = Y 85 (difference caused by inflation offset).

Fig. 9.2 shows the trend of the Japanese economy during the period of 1965–1990. It is postulated that an era of „American capitalism" is drawing to a close and another era is beginning. The 1980s are called „the decade of restructuring", characterized by:

1) Global competition.
2) Deregulation.
3) Accelerating technical change.
4) Takeover threats.
5) Restructuring of half the 500 companies.
6) Shedding business.
7) Employee lay-offs.

The principal challenges are to management innovation and technical innovation. Manufacturing firms must find ways to add value to their business.

There were three stages of structural social changes in Japan during 1965–1988:
1965–72: rapid growth; search for affluence in materials.
1972–78: stable growth; satisfaction of affluence.
1978–88: lower growth; search for mental affluence.
The Japanese chemical industry has changed course since the oil crisis of the 1970s. The industry has tried to improve productivity through the lowering of production costs and energy conservation. The diversification of new business as well as the development of high value-added products are the major driving force in the industry. Chemistry is a branch that deals with molecules and a field of technology that creates new materials. Therefore it seeks value instead of quantity, shifting weight from applied research to long-range basic research.

In the chemical industries, enterprises enjoy benefits from the declining cost of materials and energy that resulted from lower crude oil prices. Although, in the long-run, they will be greatly influenced by the establishment and operation of modern new petrochemical plants in Saudi Arabia, Canada, and Singapore, in this respect, the Japanese chemical industries must still find drastic counter measures to maintain their competitiveness in the international market. [9.4]

1) The international environment changed from an export-dependent economy (1969–1985) to a globalized economy (1986–1988) to expand domestic demands in the Japanese economy.
2) Upsurge of high-tech industry.
3) The technology transfer of commodity products to the Asian nations in the Pacific basin.

Now, in the process of the Japanese industrial structure shifting from materials to knowledge, from hardware to software, the total social and economic structure should change. The ultimate goal of management in manufacturing is not simply to produce commodities, but also to produce sophisticated value-added specialty products. We must feel the keen sense of crisis that will confront us unless management culture is not drastically changed. Now let us look at the history of the Industrial revolution from a new perspective.

The manufacturing industry initiated economic growth from its earlier position in the industrial process. As the economy expanded and society became more affluent, emphasis shifted to the finished products toward the end of the chain. These sectors then became the new leaders of industrial growth. Recently, the lead has been shifting to the service and information industries, a software-oriented sector of the economy.

In the United Kingdom, the proportion of the labor force engaged in secondary and tertiary industries surpassed those in agriculture in the early nineteenth century. In the late 1950s, the proportion of those engaged in tertiary industries surpassed those in secondary industries, marking the beginning of a trend towards a software-oriented society.

In the U.S., the agricultural labor force was surpassed by workers in secondary and tertiary industries around 1910, one hundred years later than the U.K. The labor force in secondary industries was far surpassed by those in the tertiary sector in the early 1950s, a step ahead of the U.K. The U.S. economy had begun its transition toward a thoroughly software-oriented society.

The Japanese economy reached the software age in the 1970s, twenty years behind the U.S. and the U.K. These facts suggest that, as the world approaches the twenty-first century, we are moving toward a post-industrial society. This transition is comparable with the industrial revolution of the nineteenth century in Europe and is apparent in the strength of the Japanese economy.

For example, the ratio of Japanese current account vs GNP in 1983–87 is shown as follows:

(1983) Japan   1.3 %   (1987) Japan   4.3
         U.S.   −1.4 %   U.S.   −3.3 %
In 1988, Japanese growth was 6.0 % (nominal) and 4.5 % (real).

9.3
Public and Private Cooperation in R & D Activities for Innovative Future Technology

Chemical technology supports new materials. Chemistry changes the structure of materials and creates new materials. Recently, market needs have become more

Table 9.1 Present status of Japanese economy*

	1984 Results	1985 Results	1986 Results	1987 Results	1988 Estimate
Cross national expenditure (nominal)	303,019	320,774	334,025	347,277	368,561
	6.7 %	5.9 %	4.1 %	4.0 %	0.0 %
Cross national expenditure (real)	281,256	693,332	300,877	310,571	324,811
	5.1 %	4.3 %	2.6 %	3.2 %	4.5 %
Final consumption expenditure private sector (real)	158,879	162,999	167,867	173,438	181,596
	2.6 %	2.6 %	3.0 %	3.3 %	4.6 %

* 1980 = 100 Fiscal Year Amount unit, 1 billon yen: percentage is to show comparison with the previous year.
Source: The Bank of Japan.

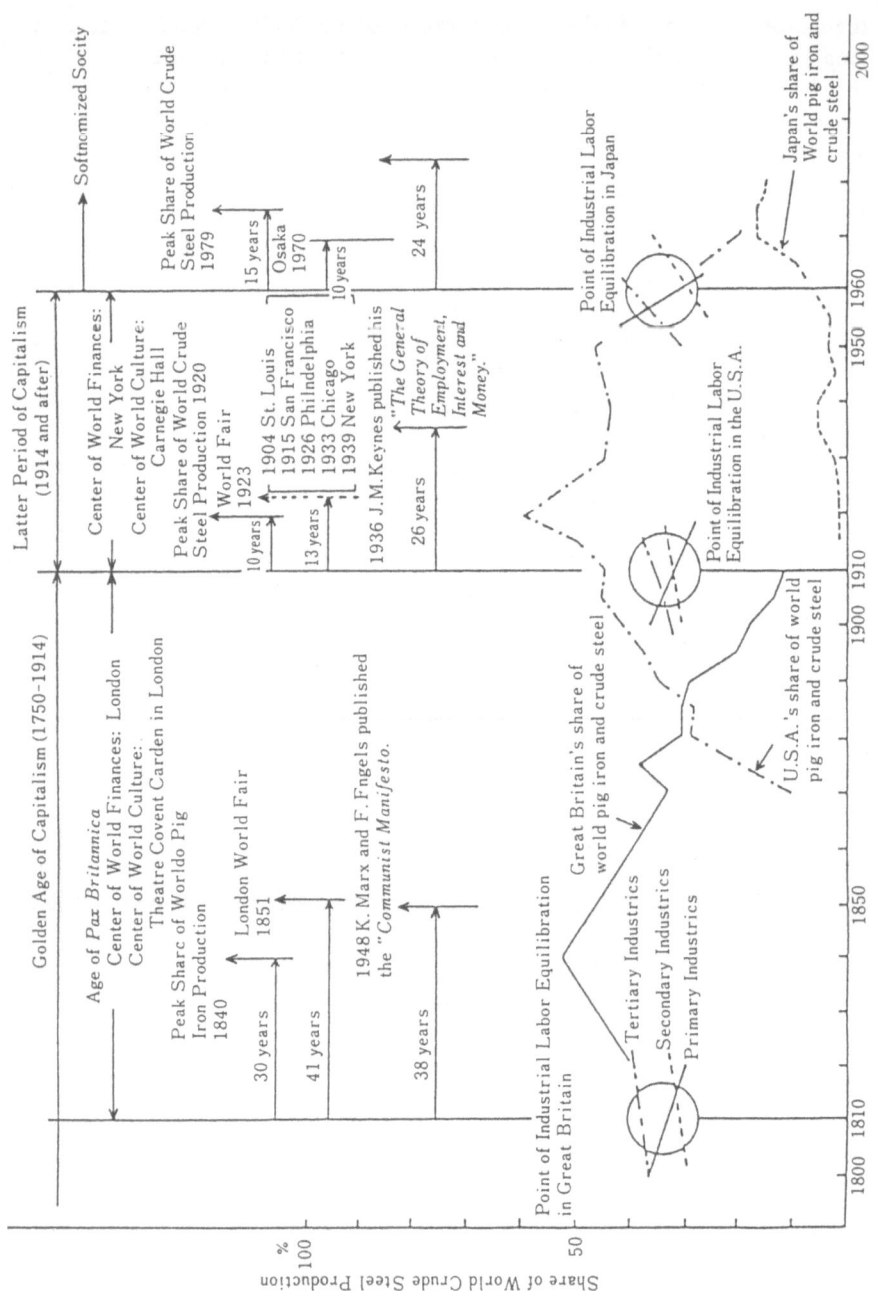

Fig. 9.3 The point of industrial labor equilibration and its historical background in Great Britain, the U.S., and Japan

Table 9.2 Trends of basic recognition and response to the new environment of chemical industry in future

(1) Lower growth of chemical industry (2) Market change in advanced country (High-tech, Specialty products) (3) Higher growth of NICS and international specialization (4) Problem of Yen appreciation and trade friction (5) Resources and energy problem (6) Environmental problem	Expectation toward Development of New Material Expectation for Life Science
↗	⇓
	Toward the New Chemical Age

precise, sophisticated, and diversified. The new materials are designed in terms of molecules (the smallest unit of material) and molecular aggregates. Chemistry is again regarded as a creative science, and the arrival of a new age of chemistry is acknowledged. As shown in Table 9.2, the age of material innovation involves social demands to pursue the creation of new materials, based on the new high technology of chemistry. The chemical industry will play a more important role in the future.

Fig. 9.6 shows the two typical approaches to research in advanced materials. Whereas convential materials are treated on the macrostructural level in mass production systems, advanced materials are based on new chemistry, which deals with the design and manipulation of molecules and molecular aggregates.

Materials are designed on the level of molecules and molecular aggregates. They should therefore be structurally controlled on the order of 0.1 microns or 100 Å. For example, one trend is to make products narrower, smaller, and very thin.

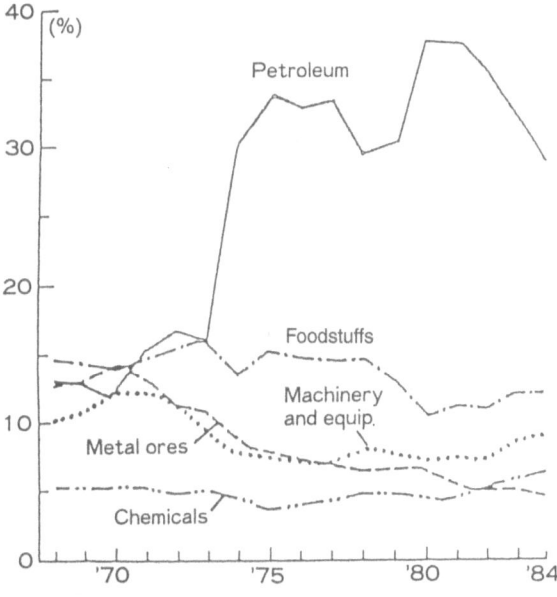

Fig. 9.4 Trends in Japan's imports share by commodity (1968–1984)

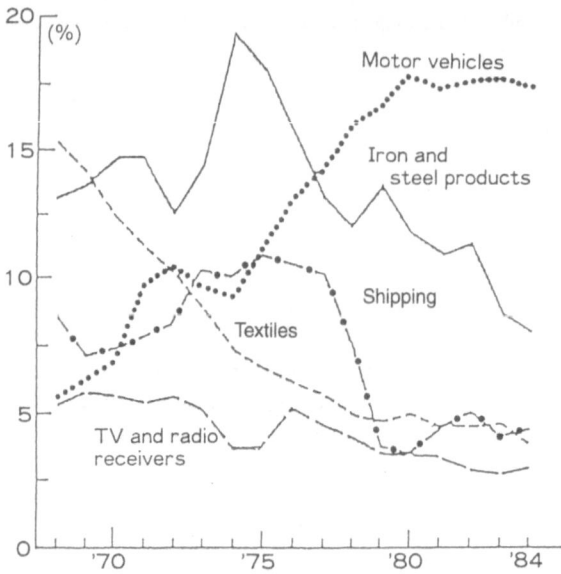

Fig. 9.5 Trends in Japan's exports share by commodity (1968–1984)

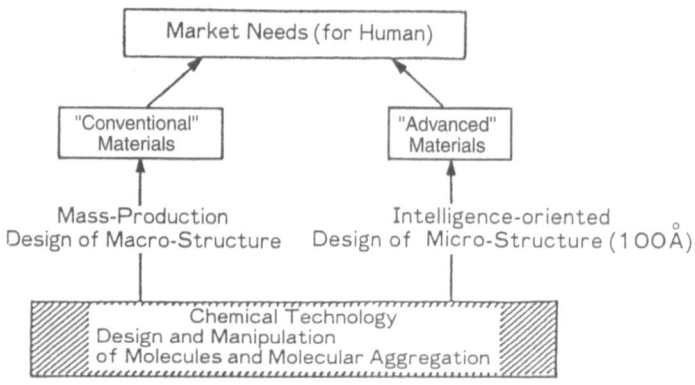

Fig. 9.6 Advanced materials

Research should be function-oriented. In other words, to meet the customer's specification, any type of new materials can be accepted, such as polymer, inorganic, or metal alloy, and composites.

The possibilities of the consumer selecting materials are expanded accordingly. First is the technical assistance to developing countries, bringing the basic chemical industry there through investment. Second comes the management coordination and coexistence on the same territory with the Pacific countries (APEC) and NIES, by creating fundamental technologies to support technological innovation in close cooperation with the users of such high technology products as information, communications, robotics, and similar materials in which Japan plays a leading role.

9.4
Research in Developed Countries

The percentage of GNP devoted to research in the leading developed countries is as follows: Japan, 1.82 %; U.S., 2.25 %; West Germany, 2.35 %; and France, 1.765 %. The percentages for government research expenditures for industry is Japan, 1.4 %; U.S., 33.1 %; U.K., 30.9 %; West Germany, 2.35 %; and France, 3.0 %.

In order to set up joint public and private research, two important factors should be considered, as shown in Fig. 9.8.

It should be noted that industrial policy (A) is driven by the initiative of the private sector. A realistic approach for Japan is informal contact between the private and public sectors, using intermediaries such as the Association of Technology Development. In 1978–79, the author assumed the post of chairman of the MITI National Project Planning Committee (GISHINKYOU), setting the research agenda for thirteen projects. Our main activities were to define the specific goal of the project, the number of research scientists, and the research budget for the decade of the 1980s. In each case, creativity of the scientists and leadership by the project manager are indispensable.

From a realistic perspective, the results of research projects differ, depending upon whether the industry concerned is strategic, important, or basic, as shown above.

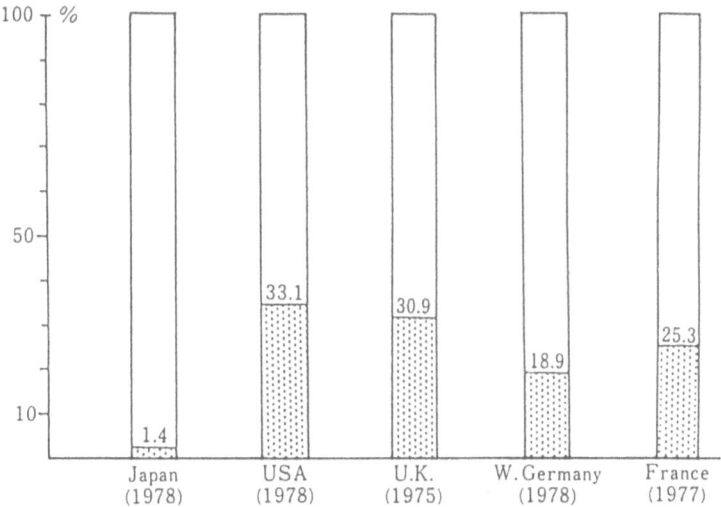

Fig. 9.7 Ratio of research expenditure born by government in the industrial field

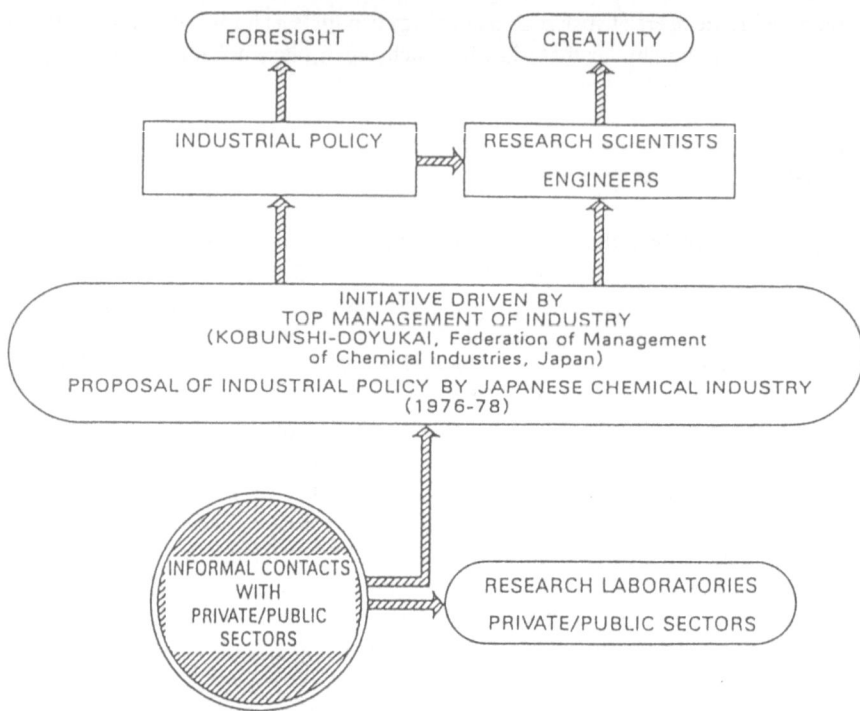

Fig. 9.8 Agenda of Japanese national R & D projects

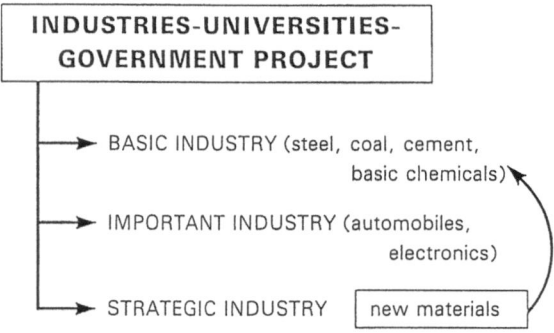

Fig. 9.9 Research projects and industry

9.5
New Advanced Materials

As to the development and marketing of new advanced materials, the main problems are:

1) Research should be original.
2) Material design should be based on market needs and should establish production technology.
3) Research takes time.
4) The production of specialty products for the international market will be smaller than that of commodity products.

To penetrate the market:

1) Materials should meet customer specifications.
2) The customer should trust the material; this is indispensable, especially for structural industrial materials.
3) The added value of the total system should be drastically increased, or the operating cost and/or production cost of the total system should be decreased.

9.6
Setting Research Agendas and Organization of the Public and Private Sectors

Kobunshi Doyukai (Federation of Management in the Polymer Industry, Japan) is a private, nationwide body that represents 108 major chemical companies in Japan. It studies domestic and international economic and technological problems, maintaining close contact with governments, research institutes, and various economic sectors at home and abroad.

The federation endeavors not only to find practical solutions to these problems, but also to contribute towards development of a sound national economy, as well the strengthening of friendly economic relations with foreign countries. It was established in April 1974 as an affiliated body of the Japanese Society of Polymer Science. It is a vital problem for Kobunshi Doyukai to strengthen economic relations with the United States, the European Community, and the rest of the world.

In the last decade, the federation experienced hardships imposed on the Japanese chemical industry by the oil crisis. It published several reports during that period and submitted several timely proposals, including: Proposal for Industrial Policy by Japanese Technological Management (1978); Technological Management Strategy of the European and American Chemical Industries – Compared with that of Japanese Chemical Industry (1980); Technical Innovation and the Chemical Industry – Overview and Task (1982); and Natural Resources, Energy Problems of the Chemical Industry and the International Competition, with Major Focus on the Polymer Industry (1982) The chronology of submitting proposals for industrial policy by Kobunshi Doyukai during the 1977–1994 period is shown below.

The chemical industry recovered its status by gradual recovery from the recession. It may be necessary, however, to take drastic measures in the chemical feedstock and basic chemicals sector to respond to the start-up of production of petrochemicals in Saudi Arabia, Canada, and other countries that supply natural resources. In the new era of technical innovation and new materials, it is expected that the chemical industry will take a leading role.

Table **9.3** List of Activities of Private Initiative for Industry Policy (1974–1988)

	Kobunshi Doyukai founded in 1974
	(Federation of Management of Chemical Industry)
	(Division of Society of Polymer Science, Japan)
1977:	*Proposal of Industrial Policy of Japanese Chemical Industry to Keidanren* (Federation of Management, Business Round Table)
	(MITI, and Agency of Science and Technology)
1978:	Setting up the „Association of Application Technology of Polymer Industry" with 13 companies
1978:	„Committee of Setting Research Agendas of National Projects" with 7 members of Private Sectors
	(Chairman, N. Yoda) Recommendation of 10 projects: (MITI-Agency of Technology Development)
	(Budget of MITI for Future Technology)
1979:	Economic Study Mission of Management strategy of Chemical Industry in U.S. and Europe
1980:	Report on the Industrial Policy of Japanese Chemical Industry, comparison with U.S. and Europe
1981:	„Research and Development Projects of Basic Technology for Future Industries" for 10 years
	(New Materials, Biotechnology, New Electronic Devices)
1984:	Second Economic Mission to U.S. and Europe
1985:	Report on the Management Strategy of Chemical Industry in New Chemical Age
1987:	First International Congress of „Association for the Progress of New Chemistry" (Fund: 4.2 b. Yen)
1988:	Fifth International Symposium of R & D Projects of Basic Technology for Future Industries
	– New Material and Innovation Technology –
1988:	*New Scheme of R & D Organization of Industrial Technology for International Joint Research (Fund: 3.9 b. Yen)*

Under these circumstances, Kobunshi Doyukai sent economic missions teams to Europe and the United States in 1979, 1984, 1988 and 1993. It also submitted practical proposals to the Japanese Government and the Business Round Table (Keidanren) on how to cope with future challenges to the chemical industry.

To illustrate these problems, one of these changes involved polymers. Among the new materials, polymers are expected to be versatile and easily processed. The living cell is made of only 20 amino acids. It is possible to obtain desired materials by combining the necessary molecules and by the polymerization process. The expected characteristics and properties of polymers are as follows:

1) high strength, high modulus
2) high flux and separation
3) electronic property (superconductivity, semi-conductivity, induction property)
4) optical property (opto-reactivity, optomagnetic property, opto-nonlinear property)

The challenge was to obtain these characteristics. This agenda is being successfully carried out by private and public joint national research projects under the System for Basic Technology for Future Industry. Some of the results were reported at the International Symposium held in Kobe on March 22–25, 1988.

As a result of national projects, 144 patents were issued within seven years; 161 papers were published in academic journals and presented at academic conferences around the world.

The tables below provide detailed information about the project.

9.7
Prospects for the Chemical Industry in a New Chemical Age

Strategic management of new materials development has recently been recognized as one of the most crucial issues for long-range planning. Based on my experience with the R & D activities of Toray Industries, I think that the key issues for the development of new advanced materials are listed below.

1) Strong emphasis on innovative and original basic research. For long-range research on basic technology for the next generation of industries, private-public-academic cooperation is desirable.

2) Since the development of new materials is highly risky, the top executive himself should patiently support R & D. Although investment in hardware was important in developing conventional materials, it is even more important in the development of new materials for top management to make decisions to invest in software.

Table 9.4 Overall view of agendas in the project

New Materials
- High performance ceramics
- Synthetic membranes for new separation technology
- Synthetic metals
- High performance plastics
- Advanced alloys with controlled crystalline structures
- Advanced composite materials
- Photo-active materials

Biotechnology
- Bioreactor
- Large scale cell cultivation
- Utilizing recombinant DNA

New Electronic Devices
- Superlattice devices
- Three dimentional ICs
- Fortified ICs for extreme conditions
- Bio-electronic devices

Research and Development
Project of Basic Technology
for Future Industries

Table 9.5 Research and development project of basic technology for future industries (jisedai)
– The objectives –

1) Promoting highly innovative R & D themes with widespread application potential as a leading technology-based country.
2) Conducting R & D activities on the themes of high risk and vast expense by collaboration of industry, academia and the government.
3) Funding R & D of the research consortia of private enterprizes to utilize their expertise, with parallel activities at the national laboratories.
4) Nurturing technologies of industrial importance from the embryotic to pre-commercial stage.

Table 9.6 Research themes and organization

New materials	
Research themes	Implementing organizations of private sector
High performance ceramics	Engineering research association for high performance ceramics
Synthetic membranes for new separation technology	
Synthetic metals	Research association for basic polymer technology
High performance plastics	
Advanced alloys with controlled crystalline structure advanced composite materials	Research and development institute for metal and composites for future industries

Table 9.7 Project name and Private/Public sectors

No.	Project name	Private sector	Public sector
1	High Performance Ceramics	Engineering Research Association for High Performance Ceramics	Government Industrial Research Institute, Nagoya Government Industrial Research Institute, Osaka Mechanical Engineering Laboratory National Institute for Research in Inorganic Materials
2	Synthetic Membranes for New Separation Technology	Research Association of Polymer Basic Technology	National Chemical Laboratory of Industry Industrial Products Research Institute Research Institute for Polymers and Textiles
3	Synthetic Metals	(the same)	Electrotechnical Laboratory Research Institute for Polymers and Textiles
4	High Performance Plastics	(the same)	Research Institute for Polymers and Textiles

Table 9.7 continued

No.	Project name	Private sector	Public sector
5	Advanced Alloys with Controlled Crystalline Structures	Research and Development Institute for Metals and Composites for Future Industries	Mechanical Engineering Laboratory Government Industrial Research Institute, Nagoya National Institute for Metals
6	Advanced Composite Materials	(the same)	Industrial Products Research Institute Mechanical Engineering Laboratory Research Institute for Polymers and Textiles Government Industrial Research Institute, Osaka
7	Bioreactors	Research Association for Biotechnology	Fermentation Research Institute Research Institute for Polymers and Textiles National Chemical Laboratory of Industry
8	Large Scale Cell Cultivation	(the same)	–
9	Utilizing Recombinant DNA	(the same)	Fermentation Research Institute Research Institute for Polymers and Textiles National Chemical Laboratory of Industry
10	Superlattice Devices	Research and Development Association for Future Electron Devices	Electrotechnical Laboratory
11	Three Dimensional ICs	(the same)	(the same)
12	Reinforced IC's for Extreme Conditions	(the same)	(the same)

3) Open-minded joint work will make new materials more acceptable to potential customers. Official agencies like MITI can promote new materials by testing them in national projects and showing them at trade fairs.
4) Joint research projects involving customers who trust each other lead to confidence in the usefulness of new materials.

In 1987, the MITI Committee for the study of Industrial Basic Technology in the twenty-first century selected the research subjects suitable for national projects, as summarized below.
An international project for the Human Frontier Science Program was proposed by the Japanese Government at the Venice summit meeting in June 1987 In the past several years, Japanese scientists made great efforts to develop composite industrial materials for future industries. For example, R & D for high performance wire preforms for carbon fiber and silicon carbide/aluminum fiber reinforced metal (FRM) composites is being carried out by university-industry-government institutes.
Recent results show that high performance wire preforms for nicalon reinforced

Table 9.8 Research program

	Phase I ('81–'84)	Phase II ('85–'88)	Phase III ('89–'90)
Synthetic Membranes for New Separation Technology	Clarification of separation mechanisms of membranes. • Primary and higher structure • Polymer functionality for permselectivity Screening of polymers	Development of new membrane materials and structures. Development of new design of modules and separation systems.	Fabrication of high performance membranes and systems. Scaled up evaluation.
Synthetic Metals	Elucidation of mechanisms of electric conductivity of: • Polyacetylenes and other linear conjugated double bond polymers. • Aromatic and heterocyclic one dimensional conductive polymers. • Chalcogenides • Graphite • Complex compounds Screening of polymers.	Molecular and structural design and synthesis of organic materials having high electric conductivity.	Fabrication of high-electro-conductive polymers into electric and electronic devices. Scaled up evaluation.
High Performance Plastics	Molecular and higher structural design for stiff and high-strength polymers. • Design for novel polymers of high strength • New methods of orientation and crystallization of molecules • Control of molecular length Screening of polymers.	Molecular and structural design and synthesis of new polymers of very high strength.	Fabrication of high performance polymers into shaped goods. Scaled up evaluation.

Table 9.9 Project budgets for 1981–1986 (million Yen)

	1981	1982	1983	1984	1985	1986	1987
New Material	1 356	2 596	3 191	3 258	3 523	3 572	3 538
Biotechnology	675	1 043	1 191	1 201	1 252	1 220	1 085
New Electronic Devices	673	1 128	1 451	1 478	1 585	1 542	1 404
Total (each year)	2 714	4 786	5 850	5 952	6 445	6 350	6 043
(US $) million Dollars Conv. Rate (130 Yen/$)	20.9	36.8	45.0	45.8	49.6	48.8	46.5
Grand Total (1981–87)		38 140 million Yen 293 million Dollars			(–1990) 50 000 million Yen 385 million Dollars		

In 1987, R & D Budget of National Project of *Separation membranes of alcohol separation* is allocated in the amount of *131 mil. Yen* as Special Account of Alcohol Monopoly Business.

Table 9.10 Industrial Basic Technology in 21st Century

1) New materials by the precise polymerizations
2) Super-conductive machine at high temp.
3) Non-linear opto-electronic materials
4) Super magnetic material
5) Molecular further material
6) Material design with precise polymerization
7) Biocompatible material
8) Advanced cupesite material supre environmental resistance
9) New metal alloy inter-metal agenda
10) Fine ceramics

aluminum matrix composition materials are successfully obtained by joint research. This is an example of successful research collaboration performed as part of the national project under the management of the Research and Development Institute of Metal and Composite for Future Industries.

The study elucidated the microstructural evolution and mechanical property changes during the fabrication process of FRM composites and in service degradations, including nuclear reactor environments.

In Fig. 9.10, economic growth and the contribution of technological progress are expressed by the following equation:

Rate of technology progress equals the growth rate of the economy minus the growth rate of capital minus the growth rate of labor.

The new advanced materials are expected to satisfy future human needs, and the chemical industry should respond to these growing expectations.

In Fig. 9.11, the market size for new materials is projected to grow from half a billion yen to 10.2 trillion yen in 2000.

The annual growth rate of plant investment was greater than 10 % from 1988–1990, which was the highest since the high growth period (1965–1970) in Fig. 9.12. This

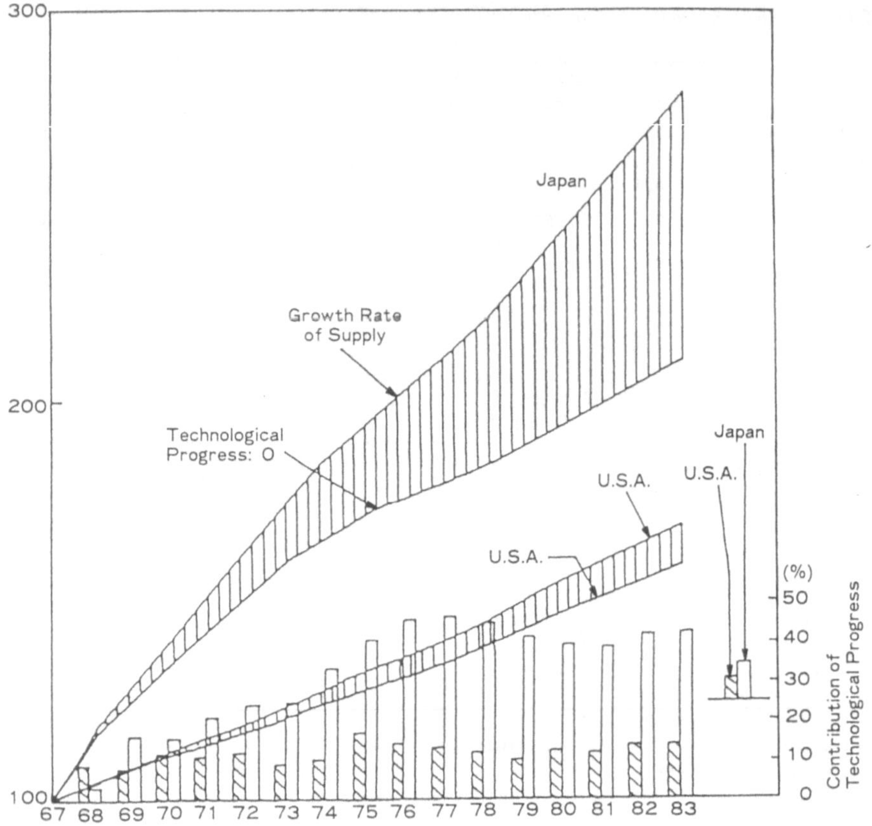

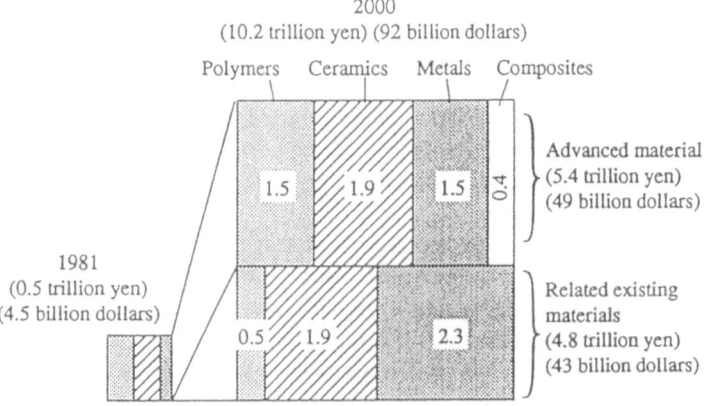

Fig. 9.10 Economic growth and the contribution of technological progress
Source: "White Paper of Trade" issued by MITI (1985)

TECHNOLOGY MANAGEMENT AND GLOBALIZATION

Fig. 9.11 New material market expected in chemical industry in the year 2000 (Total market size in Japan: 10.2 trillion yen)

Table 9.11 Scheme for „Basic Technology Promotion Center" (Approved in 1988)

Investments (NTT stock profit) Financing without Interest Private Expenses		Investments (3.9 b. Yen) Subsidies (special accounts) Private Investments
R & D Infrastructure Basic Planning Project	R & D PROJECTS CORPORATION	INTERNATIONAL R & D Joint Projects

Fee for use — Planning Operation — Contract research — Commission-contract — Subsidy

Large Scale Research Facilities Data Base etc.	R & D ASSOC. OF PRIVATE CORPORATIONS	Res. Scientists entry	INT´L JOINT R & D

Use — Joint Research — Information exchange — Entry for research

Private Companies (CUSTOMERS)	Government Research Institute (MITI)		Foreign Research Scientists

Research Scientists of Private Sectors and Gov./Academic Institutions

increase in plant investment was not only due to cyclical economic change but also to such independent factors as deregulation, internationalization, labor-saving measures to cope with labor shortages, technical innovation, and the revitalization of research and development that reflects innovation. The analysis of technological innovation and corporate research and development activities shows that they are closely related to each other.

On an international level, if research efficiency can be indicated by the number of patent applications per scientist, Fig. 9.13 suggests that recent performance of Japanese scientists has been outstanding. As can be seen from Fig. 9.13, Japan's technological power has gradually surpassed that of other countries. In the past, we depended on the introduction of technology from overseas but, as indicated by the graph, this situation is changing. Japan now depends on itself for new technology.

Table 9.12 Scheme of R & D organization of industrial technology projects of MITI in 1988 (yen in billions; dollars in millions)

Research Center for non-gravity Environmental experiments (Hokkaido-Coal Mine)	Oceanographic bio material Research Center (Kamaishi/Shimizu Biotechnology)	Ion beam for new material Modification of materials (Kansai Academic Center)	
5.2 yen (4 dollars)	6.0 yen (5 dollars)	7.8 yen (6 dollars)	
Gov.	2.65 yen	5.0 yen	
Dev. Bank Financing	+ 1.73 (local)	–	+ 3.1 yen
Private	0.87 yen	1.0 yen	2.0 yen

Total 19 billion yen (15 million dollars)

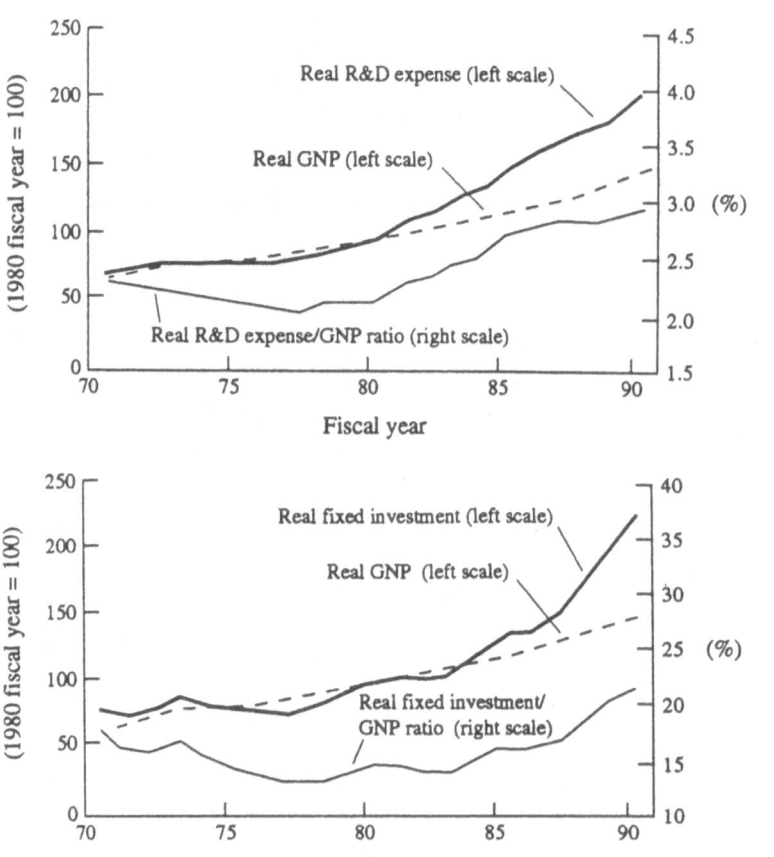

Fig. 9.12 Research and development and fixed investment
Source: Economic Planning Agency, the Japanese Government (1992)

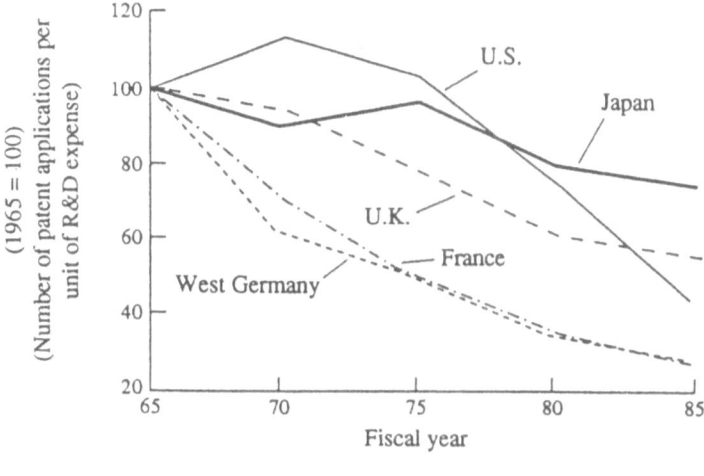

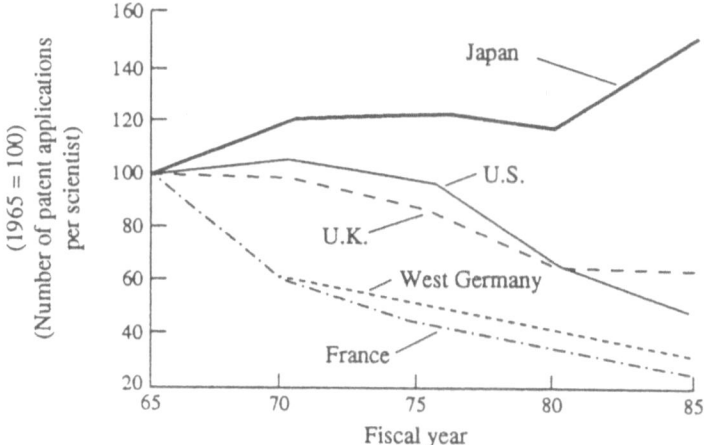

Fig. 9.13 International comparison of research and development efficiency
Source: OECD (1987)

Technological innovation and economic growth is focused on the contribution of technological innovation to economic growth. Research and development activities bring technological innovation, which in turn acts to support investment in the plant. On the supply side, the productivity increase brought about by plant investment will continue to contribute to Japan's macro growth. The growth trend of our manufacturing industry during the last twenty years shows greatest contribution from research and development-intensive industries: Electronics, the chemical industry, and precision machinery. These three industries currently account for as much as one-third of the real GDP of total manufacturing, up from only about 10 % in 1970. Also, an analysis of the growth of manufacturing industry in terms of capital, labor, and

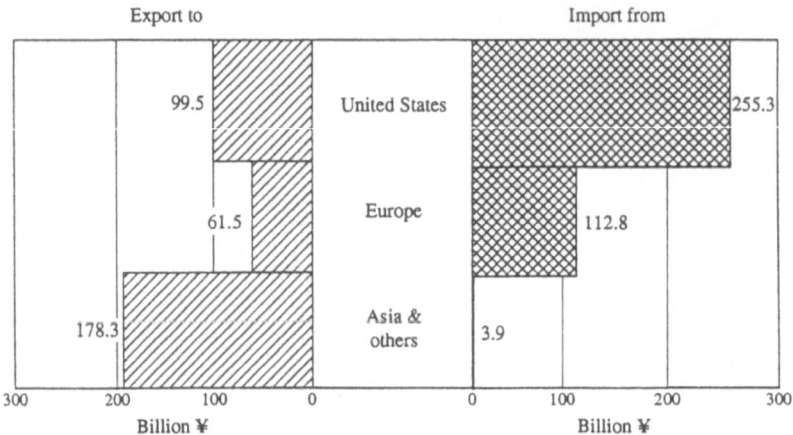

Fig. 9.14 The Japanese technology trade in 1990
Source: Statistic Bureau

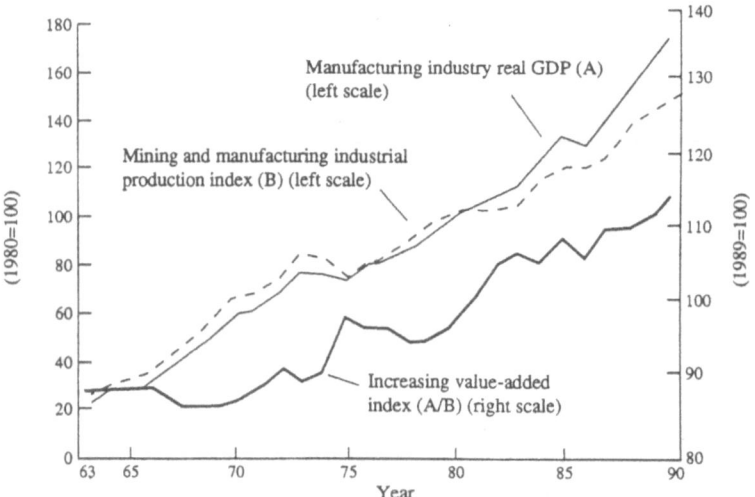

Fig. 9.15 Increasing value-added in the economy
Source: Economic Planning Agency, the Japanese Government (1992)

technological advancement indicates that the contribution of technological advancement to GDP growth was decreasing in the 1970s, but started to increase in the early 1980s as shown in Fig. 9.15. This reflects the advancement in technological innovation that started around 1980. Considering that the increase in productivity during the 1960s was dependent on imported technology, Japan's recent productivity increase is noteworthy for having been achieved mainly through research and development efforts.

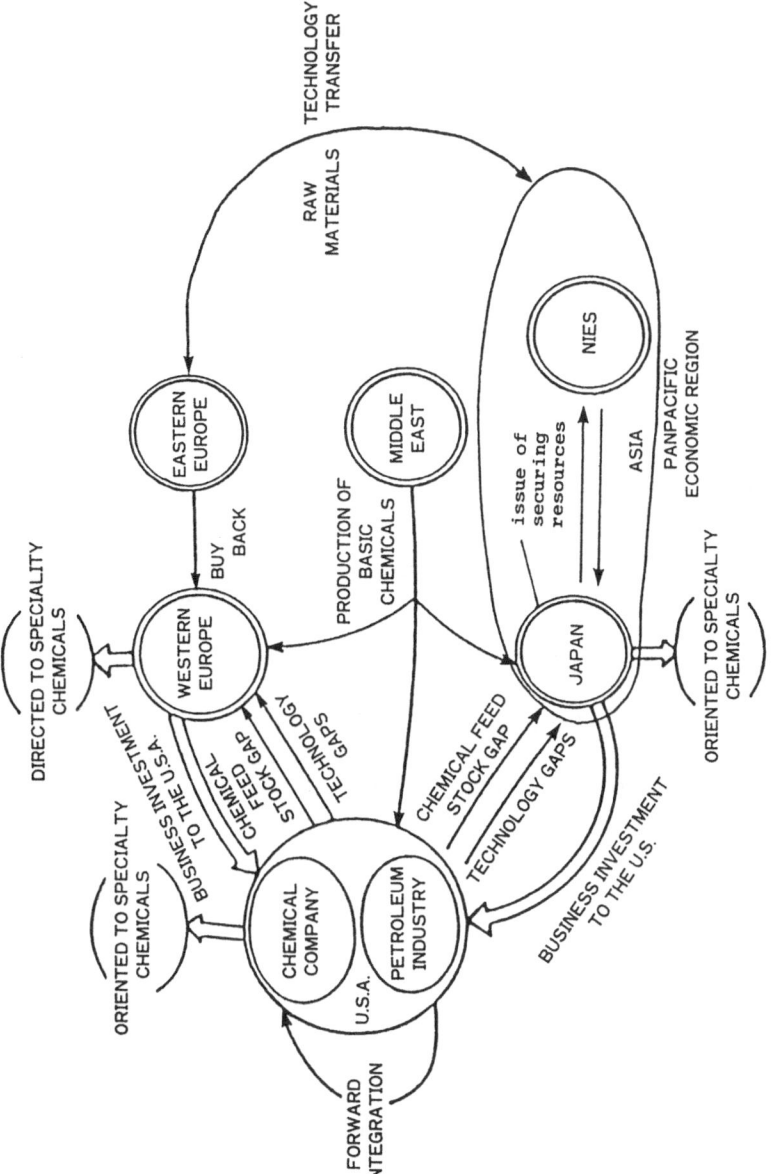

Fig. 9.16 Correlationship of economic flows of world chemical industry (1995)

The Japanese technology trade balance of exports and imports in 1990 is shown in Fig. 9.14. The deficit of technology trade between Japan and the U.S.A. is obvious, and the imbalance should be improved in the future. [9.8]

As to the high value-added and qualitative economic changes in the manufacturing industry, Japanese manufacturers have become very active in developing and producing higher value-added products that increase in value per unit of production, as shown in Fig. 9.8. A value-added index is used, which is equivalent to the value-added per production unit, calculated by dividing real GDP, which represents the total value-added, by the industrial production index, which is based on the production of different products measured by such units such as tons and numbers. As changes in the value-added index are shown in Fig. 9.15, the index increases rapidly around 1980, as was the case for other indexes. The rapid technological innovation occurred in this period and played an important role in this phenomenon. [9.9–9.14]

9.8
Conclusion

Finally, the author wish to point out the importance of both the Pacific Basin economy and the relationships between the U.S. and Japan, and between Japan and the European Community respectively as shown in Fig. 9.16.

The area of the Japanese islands is only 0.25 % of the total area of the globe, and only 2.5 % of the world's population lives there. Japan annually imports 600 million tons of raw materials, including iron ore, coal, and crude oil (200 million tons). Japan annually exports 90 million tons of value-added manufactured products. Japan's GNP represents 14 % of the total world economy. The population of Japan is one-half that of the U.S., the GNP of Japan at 360 trillion yen is one-half of the GNP of the U.S. The GNP per capita of Japan is 69 % of the U.S. figure.

9.9
International Competition and Cooperation

1) To be internationally competitive, industries must give high priority to investment, to modernization leading to improved productivity, and to R & D pointing both towards new products to meet changing market needs and to new process concepts. Industries must also recognize the need to be competitive in worldwide markets.
2) Governments – individually and collectively – must adopt policies which provide a favorable climate for industrial growth and competitiveness, especially support for R & D and education.
3) In the last decade, the worldwide chemical industry has faced a number of critical challenges – including interruptions of oil supplies, a dramatic escalation in feedstock prices, and serious inflation followed by a worldwide recession. In addition, Middle Eastern oilproducing countries are beginning to move into the petrochemical business, promising increased competition down the road.

The chemical industry has met these challenges in a variety of ways. Most major Japanese companies have substantially reduced their energy consumption per unit of production and have improved product yields from petrochemical feedstocks. Perhaps the most important long-term response to the challenges of the past decade, however, is a renewed commitment to research and development. [9.6–9.9]

While much can be done in Japan, other leading economic powers must also play a global role. Japan is the free world's second largest economy; its gross national product in 1987 was about 70 % larger than that of the Federal Republic of Germany, the world's third largest economy. Japan is the world's third largest exporter.

Chemistry will play a significant role in creating new materials, because it manipulates and creates new substances on the molecular level. The twenty-first century is regarded as the new chemical age. Based on my close working relations with MITI and the Society of the Japanese Chemical Industry, the following points are important for developing new technology and advanced materials: [9.10–9.16]

1) Strong emphasis on basic innovative research through private initiatives and joint work with government / academia / industry.
2) Direct risk-taking by corporate executives.
3) Development of joint work with government / academia to create new materials.

The new advanced materials are expected to satisfy future human needs, and the chemical industry should respond to these growing expectations. Japan and the United States can make significant contributions to this effort by resolving their mutual conflicts. As the free world's leading economic nations, both must recognize that economic power carries with it both opportunities and responsibilities. If we seize the initiative, seek cooperation, and demonstrate wisdom, we can fulfill our responsibilities to the global community. [9.17, 9.18]

9.10
References

9.1 Yoda N (1988) Invited lecture at Seminar of MIT-Japan Science and Technology Program, 12 April, 1988 MIT, Boston, MA, U.S.A.
9.2 Yoda N (1988) Invited lecture at Conference of Amer Chem Soc on Future Opportunities for Polymers, 9 Sept, 1988, Washington, DC: Yoda N (1989) in: Alper N, Nelson G L (ed) Polymeric Materials Chemistry for Future, Chapter 3, New Frontier p 44 Am. Chem. Soc, Washington, DC
9.3 Dagani R (1988) Chem. Eng. News, 26 Sept (1988), p 27
9.4 Lepkowski W (1988) Chem. Eng. News, 7 Nov (1988), p 4
9.5 Yoda N (1988) in: A Plenary Lecture of InterChem 88 Conference on Specialty and Petroleum-basee Chemicals in Asia-Pacific Technolgy and Ind Development, 12 Dec 1988, Hongkong
9.6 Yoda N (1989) Toray Corporate Business Research, TBR Intelligence 1 : 2 p 48
9.7 Yoda N (1990) TBR Intelligence 2 : 2 p 56
9.8 Yoda N (1990) TBR Intelligence 3 : 1 p 23; Yoda N (1991) TBR Intelligence 3 : 1 p 38; Yoda N (1991) TBR Intelligence 3 : 2 p 16; Yoda N (1992) TBR Intelligence 4 : 1 p 15; Yoda N (1992) TBR Intelligence 5 : 1 p 28
9.9 Yoda N (1990) in: Management Strategy for Competitiveness and Cooperation in Borderless Economy, Am Chem Soc Mini Symposium, 27 August, 1990, Washington DC

9.10 Yoda N, A Plenary Lecture of IOPAC 94 Conference on Speciality and Petroleum-Based Chemicals in Asia-Pacific Technology and Industrial Development, November 1994, Taibei, Taiwan

9.11 Yoda N, TBR Intelligence 2, 81 (1989)

9.12 Dertouzos M L, MIT Commission of Industrial Performance, Made In America. Regaining the Productive Age, MIT Press, Boston, MA (1989); Japanese Translation, Made in America, Yoda N, Soshisha Publishing Co, Tokyo (1990)

9.13 Yoda N, paper presented to the IVA Conference on the Introduction of New Materials in Industry, Stockholm, October (1990)

9.14 Maeda E, TBR Intelligence 5, 14 (1992)

9.15 Yoda N, Japan, Intelligence Overcomes Natural Resources. Do Not Be Arrogant, Keizaikai Publishing Company, Tokyo (1990)

9.16 Yoda N, in: The Macromolecular Concept and Strategy for Humanity in Science, Technology and Industry (Ito Y, Okamura S, and Råndby B, Eds.), Springer Verlag, Heiderberg, Chapter 5 and 6 (1996) in press

9.17 Yoda N, J. Polym. Sc. Polym. Symp. 75, 125 (1993)

9.18 Yoda N, Prog. Polym Symp. 75, 975–994 (1994)

Chapter 10

Management Strategy for R & D and Industrial Policy

YOSHIKAZU ITO

Abstracts

In today's chemical industry, research and development – the process of seeking out new technology for the twenty-first century – has become an integral part of management strategy. Indeed, promoting new business on the basis of the creative R & D of individual firms is a major trend in Japan. Therefore, the most effective managers will be those who train and motivate competent researchers and technocrats, and give them opportunities to exploit their capabilities. The Japanese chemical industry would benefit greatly if joint national research projects were arranged on the basis of cooperation among government, industry, and academia. This is also true for the Chemical Society of Japan. Everything possible should be done to: (1) deepen relationships among the Federation of Technological Management of the Chemical Industry, the Chemical Society of Japan, and other professional associations; (2) strengthen ties between government and industry; and (3) raise the level of cooperative efforts between industry and educational institutions.

10.1
Introduction

The author wish to mark the third anniversary of Toray Corporate Business Research, Inc. be congratulating the unique thinktank, established in July 1986 on its success. Its objectives and tasks are getting more important for the globalization of management strategy.

In March 1988, the author became president of the Chemical Society of Japan. To convey some notion of the dimensions of the organization, about 7 800 people participated in its annual meeting in 1988; about 3 800 papers were submitted to the session, and 124 speakers made presentations. These figures are much higher than those for earlier years. This fine group has done much for the Japanese polymer industry, especially by joint national research projects carried out by universities and industrial firms. For example, the author has been privileged to have served as chairman of the Research Project for Basic Polymer Technology, which is supported by the Ministry of International Trade and Industry. In this capacity, the author has observed the work of ambitious young researchers for the project and noted how the combined effects of

Okamura, Rånby, Ito (Eds.): Macromolecular Concept and Strategy
© Springer-Verlag Berlin · Heidelberg 1996

competition and cooperation stimulate rival companies in the industry. When such programs are effectivly managed, they produce significant results. At Toray, my policy is to stress the need for both basic research within the company and participation in national basic research projects. If we Japanese throw ourselves into this kind of work with the enthusiasm that it deserves, the author is sure that the resulting achievements will meet the highest international standards. The author has become convinced that research that is useful for business can be performed effectively in joint facilities that operate under a new concept of cooperative competition in which personnel are provided by several companies. In such a setting, researchers can stimulate each other while learning from each other, competing as they cooperate. When people are exposed to new influences in this way, rather than being confined to the limitations of one company's basic research program, the payoff in motivation and creativity can be tremendous.

In pursuit of this kind of creative interplay, the author feels that the Society of Polymer Science, the Chemical Society of Japan, and other professional organizations should maintain close ties that involve many industrial firms. In this way, the can pool their resources and work closely with government agencies, setting ambitious long-term goals, agreeing on research themes, and working together to fulfill them. No other strategy can enable the Japanese chemical industry to compete successfully with its counterparts in Europe and the United States.

10.2
Global Management Strategy Survey Missions in 1979, 1984, and 1988

Since 1979, the author has participated in three missions to study the chemical industries of Europe and the United States. This experience enabled me to explore some of the ideas discussed above with foreign colleagues. The author learned that the major U.S. and European chemical firms operate on a much larger scale than we do. For example, a single firm may have the R & D capacity of ten Japanese companies, thus enjoying a definite strategic advantage. The author therefore reached the conclusion that it would be extremly difficult for a single Japanese firm to establish itself in the international marketplace solely thought its own competitive efforts. The author also believes, however, that if the individual firms in the chemical industry cooperate and enlist the combined efforts of government and academia, the resultant research will equal or surpass that of European and American companies. This conviction prompted the author to propose national research projects. The result has been cooperation among individual firms, educational institutions, and government agencies; among the latter have been the Ministry of International Trade and Industry, the Ministry of Health and Welfare, the Science and Technology Agency, and the Ministry of Education. In these projects, research goals have matched high international standards, and much progress has been made towards creating a system for conducting longterm basic research. Thus far, Toray has participated in ten national projects, and many of our researchers have benefited from this experience. The author is convinced that this is the most promising way for chemical industry to attain the level of American and European counterparts.

10.3
Examples of Collaborative Industrial Research

Two research institutions that have recently been launched in Japan are good examples of the kind of approach that the author advocate. One is the Protein Engineering Research Institute that was established in March 1986. Fourteen industrial firms and one government agency – the Basic Technology Promotion Center – jointly participate in the institute. It is a joint stock company with capital of 1.4 billion yen, 30 % of which is private and the balance public. The project is scheduled to last 10 years; its research budget is 1.7 billion yen. The center has constructed a laboratory in Osaka and intensive research has been under way for two years. The author is president of the center, and the laboratory director is Professor Ikehara of Osaka University. The second example is the Biomaterial Research Laboratory that was established in February 1987 with the joint participation of the government and three private firms: Sumitomo Electric Industries, Toray Industries, and Sumitomo-Bakelite Company. The laboratory's capital is 100 million yen. Its' president is Tsuneo Nakahara, vice-president of Sumitomo Electric Industries, and its laboratory director is late Dr. Shigeyasu Kobayashi, who had been research associate at the Toray Basic Research Laboratory.

Both institutions are effectively pursuing interdisciplinary and interindustrial research programs. In order to tackle these ambitious tasks, teams combining specialists from different fields have been created that are doing work that could never be attempted by a single company or university department. For example, an ongoing large-scale project at the biomaterial laboratory involves characterizing the three-dimensional structure of the protein molecule, elucidating correlation between the fine structure and the functions of protein, and wing the results to remodel and synthesize new protein and develop the functions of protein. This project is attracting considerabel attention in Europe and the United States. For example, many Americans have recently visited the facility, and researchers from the laboratory have presented papers and lectured at international conferences in various parts of the world. Some foreign visitors have expressed interest in how the laboratory is managed, and some steps are already being taken in Europe and the United States to create the same kind of joint research group.

Proposals by myself and initiatives by the Federation of Technological Management of the Chemical Industry served this purpose in Japan, while interested firms contributed people, money, materials, and equipment, thus joining forces for a common purpose. This kind of thinking and pattern of action have become fairly well established in this country. Thus, fourteen companies – both Japanese and foreign – in such diverse fields as chemicals, computers, and biotechnology are working together with bright young university researchers to make up a total research staff of some 60 people.

This kind of cooperative effort can be accomplished relatively easily in Japan, but this seems not to be the case in Europe and the United States, where it is more customary to form inter-company research groups whose purpose is limited to a certain specific goal. Nevertheless, as the author has indicated, some Europeans and Americans are interested in emulating the Protein Engineering Research Institute.

It should be stressed that the results of joint Japanese research in the life sciences should be made available to the international community for the benefit of all mankind.

10.4
R & D is Essential for Business Success

Japan's chemical companies operate at some disadvantage, because their market is segmented into small business sectors, and individual companies must cope with keen competition. However, the R & D capabilities of each company serve to enhance its competitiveness. If this strength is coordinated, a research-promotion system can be built that is much stronger than the inhouse R & D capabilities of large U.S. and European firms. Although serious management problems arise in operating an organization involving 14 different firms, it is also true that researchers from these firms and from universities express their enthusiasm by working on weekends and otherwise putting their professional concerns ahead of their private pursuits. Their example has convinced me that young researchers can achieve tremendous results if they are provided with an environment that is conductive to creative work. The author expects the academic and industrial sectors to propose a variety of research themes, and I believe that it will be necessary, in order to build an effective research organization, to eliminate barriers that separate academic disciplines and industrial sectors from one another. This is surely the best way to promote truly advanced technology. In pursuing this progressive kind of research environment, the Federation of Technological Management of the Chemical Industry has an important role to play.

These efforts will take time. The important thing is to motivate ourselves to push forward toward the goal that the author is advocating. In this connection, the author urges that each company in the federation designate one relatively young business manager with real promise as a member of the federation. My goal is for the federation to help train managers who will be capable of managing chemical firms within five to ten years. On the fifteenth anniversary of the federation next year, the author hopes that it will take on an enhanced sense of purpose.

The most recent phase of the business cycle in Japan, in 1988, occurred during a major structural transition generated by the rapid appreciation of the yen; it differs from similar phases in the past. While domestic demand is expanding, external demand is falling off. The increase in the international value of the yen has stabilized domestic commodity prices and increased real income. Changes in demand have had a dramatic effect on the industrial structure. Thus, the non-manufacturing industrial sector performed well, buoyed by the stronger yen and growing domestic demand. Meanwhile, however, the manufacturing sector suffered from the drop in external demand, and this gave rise to so-called „two-facedness" in business.

The rapid appreciation of the yen since the fall of 1985 has thus brought about a farreaching structural adjustment of the Japanese economy, and further arduous transitions lie ahead. However, such ordeals brought about favorable consequences, internally and externally, in 1988. To be specific, it looks as though fundamental conditions are being created that will result in the fruits of economic growth being seen in higher living standards for the Japanese people.

New types of economic evolution are emerging in response to such global problems as energy, resources, food, and the environment, as well as to value shifts from the material toward the spiritual. In this context, one can foresee dramatic technological developments in information, new materials, innovative energy sources, and biotechnology. These trends in turn are nurturing the growth of so-called high-tech industries, which create new business opportunities for the chemical industry. Indeed, chemical technology may be a driving force in the evolving technological revolution. The potential exists for the chemical industry to secure a profit base for the coming decades by outgrowing the traditional materials industry and evolving into a new type of technology-intensive industry.

Although chemical technology and the new materials it develops are essential for the creation of high-tech industries, the profitability of the chemical industry may nevertheless not be satisfactory as long as it remains a mere supplier of materials. Its profitability will probably depend on whether it makes effective use of its own technical expertise and on how far it manages to extend itself into downstream sectors. Strategic undertakings of this kind are strongly oriented toward R & D and require enormous amounts of time and resources. In this way, technological management strategy becomes crucial in such areas as selecting R & D themes, the efficiency of R & D systems, the optimal allocation of resources; and restructuring business to match the pace of R & D. Finally, new ways of international collaboration must be sought out that will promote global economic and technological progress.

10.5
Economic Symbiosis: Building New World Order and Concepts Used to Build Japanese Pagodas

Explanations for the strength and resiliency of the Japanese economy lie in our traditional culture. Thus, wisdom passed on by the master woodworkers who built pagodas long ago in Japan is still relevant for contemporary managers. The adages of these craftsmen counsel us to appreciate differences among individuals and to cultivate the needs of individual consumers. In the emerging borderless global society, Japan should use symbiosis to build a mutually dependent new world order. Nowadays, everything, including politics, society, economics, and technology, is changing rapidly, as paradigms keep shifting. Economic difficulties and technological innovations have greatly altered the structure of industry during the past four decades; this has led to tremendous changes in values and lifestyles. In particular, Japanese industry has been buffeted by a series of recessions and other setbacks, such as oil crises, trade frictions, and appreciation of the yen. All nations are interdependent in the borderless global economy, and we now face global environmental issues that must be addressed, as recently discussed at the Earth Summit in Brazil.

After each setback, Japanese industry promptly recovered its footing and became even stronger. What accounts for this strength and resiliency? I think the answer lies in the traditions emerging from 2000 years of Japanese history and culture, notably

Fig. 10.1 Pagodas of HORYUJI Temple (Nava, Japan, built in 607 A. D.)
Source: The MAINICHI PRESS, Japan

the network system that still operates in Japanese industry. Some of these traits can be summarized as follows.

1. Workers are diligent and the work ethic penetrates society. All company employees, from top management to workers, are well-informed about company policies and share the same goals. Suggestions by workers and their efforts to attain corporate goals are respected.
2. Customer-oriented marketing based on traditional values is given top priority, with emphasis on thorough analysis of the needs and desires of customers.
3. Japanese-style total quality control and production systems, which combine traditional Japanese methods and mass-production technology from the U.S. and Europe, heads to the efficient manufacture of excellent products.

How was this accomplished? To address this issue, the author will use the example of ancient Japanese architecture as shown in Fig. 10.1. The many five-storey pagodas in Kyoto and Nara are made from Japanese cypress wood, using methods that go back thirteen hundred years to traditional concepts of Japanese shrine carpenters [10.1].

The strength and durability of the Horyuji temple, which has withstood countless earthquakes, blizzards, and typoons since the sixth century A. D. are the fruit of artisan traditions. Master carpenters left behind some traditional sayings that are relevant to the economic symbiosis (*kyosei*) of international management of the next century as illustrated in Fig. 10.2.

One adage advises: „When buying wood, do not buy trees but an entire forest." To build a pagoda, a shrine carpenter needed a wide variety of wood with special characteristics which has grown in different environments for more than one thousand years.

Another saying: „Assembling a pagoda expresses the unique natural character of Japanese cypress wood." This means that every tree derives its unique characteristics from the soil and environment in which it grows for many centuries; the peculiar bent, twisted, or warped branches of these trees should be taken into account when deciding where to use the wood in a pagoda. For example, a tree which grew on the sunny south side of a mountain should be used on the south side of a pagoda; the north side of a pagoda needs northern wood. Wood grown in the wind is used for wind resistance

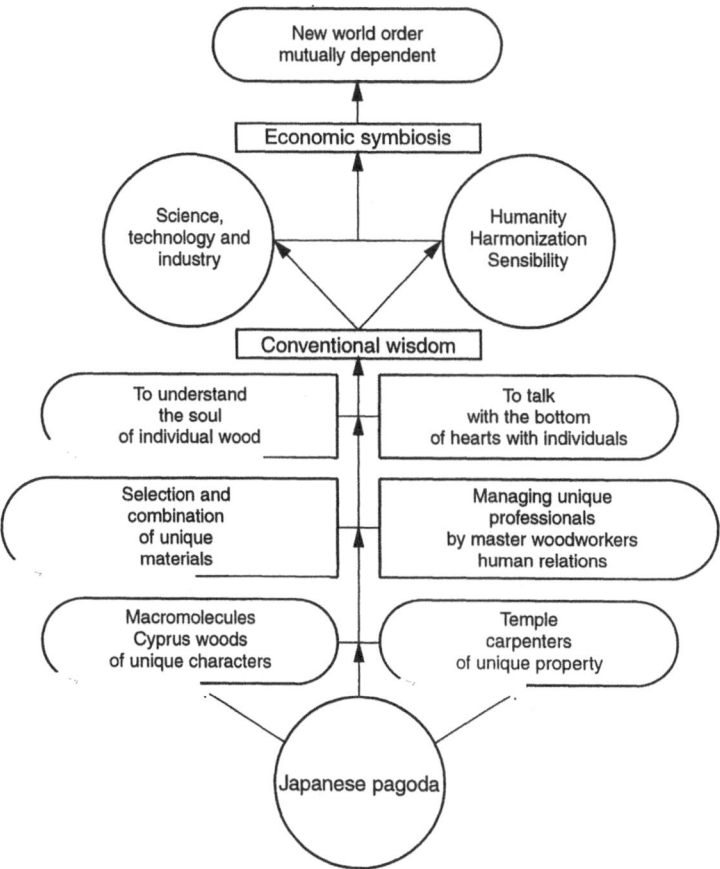

Fig. 10.2 Relationship of economic symbiosis and new world order derived from the concepts to build Japanese pagodas of conventional wisdom

at the top of the pagoda. In selecting wood, various types should be chosen that harmonize with the location of the building. Also, wood that bends to the left is balanced by wood that bends to the right.

Another adage: „In choosing a shrine carpenter, consider his personal traits and try to make them harmonize in the same way that one uses a unique piece of wood." Shrine carpenters of course had their own characteristics and habits. The chief carpenter hired young and old carpenters and gave them on-the-job training at the building site, as well as moulding them into a cooperative working team. Combining trees into a shrine is like blending men's hearts and minds into a team. Good management is indispensable to building a solid pagoda. You have to manage humans so that individual talents combine to accomplish a goal.

Still another aphorism: „Talk with each one of the trees. Trees have souls." Thus, if a shrine carpenter cannot understand the soul of trees, he cannot be a capable carpenter [10.2].

Such age-old ideas are quite different from modern management concepts that apply to mass-production systems and the prefabrication of standardized materials [10.3, 10.4]. Although, to my regret, these traditions are disappearing in Japan, they are still significant for some managers. They give us important insights into understanding how people think and how to select software and hardware.

In particular, these old ideas help us to appreciate subtle differences between individual entities and to value individual traits, human as well as material. I believe that respect for individuality will eventually create a new culture and that a new market for products catering to individuality will arise. Mass production of standardized products cannot fully satisfy future markets and the needs of consumers. We must all pay careful attention to subtle, individualistic needs if we are to become a dominant market leader.

Differences between U.S. and Japanese management styles have attracted much attention. Solutions to U.S.-Japanese trade frictions have been discussed at a recent conference in Tokyo. In my view, there are no fundamental differences between the two in the sense that the basic foundation of all management should be respect for the individual, from top management and workers through to consumers.

Society is becoming more diverse, and lifestyles are becoming more individualistic as our superindustrial society moves toward the twenty-first century. Through paradigm shifts, pluralism is penetrating the social structures of many traditional cultures. Unique customs and manners should be harmonized with each other and coexist in the age of new materials.

In order to maintain ethnic diversity in the international community, we should live together, not in a melting pot but instead in a salad bowl where ingredients do not lose their individuality. Each member of international society should keep his unique color and shape, just as red round tomatoes, green lettuce leaves, and tubular cucumbers harmonize together in a good salad. This is the basic concept of economic symbiosis (*kyosei*), which is the ecological term for mutual accomodation of different species. Kyosei involves three concepts:

1. coexistence in the domestic society and economy;
2. coexistence in the international market and society; and
3. coexistence in the global environment.

10.6
Conclusion

The concept of symbiosis uses its own strength and vitality to harmonize with the environment of other species. Each nation should respect other nations, to form a mutually dependent new world order among 180 countries.

The Japanese monoculture has two thousand years of tradition. We should now change our paradigm and try to live as a member of the international community in one interdependent global family. We live in a society of economic symbiosis (*kyosei*), where every nation is an equal independent partner. Seen in this light, the way pagodas were built 1300 years ago can be applied to the emerging international community. Indeed, I firmly believe that the methods used by Japanese shrine carpenters can be applied by managers who are building the new world order.

10.7
References

10.1 Ito Y (1991) Toray Corporate Business Research, TBR Intelligence 2 (1) : 4
10.2 Ito Y (1992). In: TBR Intelligence 5 (1) : 4; Ito Y (1992) Paper at the United States-Japan Business Directors Conference, 9 April 1992. Washington, D.C.
10.3 Nishioka J, Obara J (1985) Wood-supported Horyuuji Temple. NHK Broadcasting Corp, Tokyo
10.4 Nishioka J (1985) Learn from the wood, Shogakukan, Tokyo

Chapter 11

Polymers as the Basis of Human Body and Mind

SEIZO OKAMURA

11.1
Introduction

Historically, there are three classes of macromolecules that occupied and now occupy the focus of our interest: first, natural polymers (non-living); second, synthetic polymers; and third, biological polymers (living). From the author's viewpoint of axes of existence (see Chap. 4), polymers may occupy, firstly, the spatial axis (for non-living or passive materials), secondly, the space + time axes (for living or active materials), and thirdly, the space + time + memory axes (for human being).

Thus, the macromolecular concept in our meaning may run from the human body (as materials) to the human mind (as brain). In general, the body is considered to be separated from the mind, as proposed by De Carte 380 years ago. Apart from this body-mind problem, human beings continued making materialistic efforts that have eventually led to the current advanced technology. However, it may also be said that the relation between human body and mind corresponds to the relation between technique and arts discussed in Chap. 4. In some sense, such relations would be directional, as in the approach from technique to arts (see also Chap. 4), or analogously, from materials to phenomena and from a body-oriented to a mind-oriented way of thinking, for all of which our macromolecular concept is relevant.

In this chapter, we will discuss the body-mind relation further relative to the macromolecular concept, in an attempt to develop strategies of scientific procedures like research and development. The discussion will be directed first to the relationship between knowledge and wisdom, secondly to that between physical and psychological sciences, and finally to the human spirit. The last part of this chapter gives the tentative conclusions of this book.

11.2
Knowledge and Wisdom

Knowledge is, in general, obtained through the organs of our body that are responsible for a human's five senses (seeing, hearing, etc.) and serves as one of the elements of pure science. In contrast, wisdom is based on knowledge *and* experience (or experiments); in other words, wisdom may be obtained through not only the five senses but also through the sixth sense („sensus communis"). In each process to gain

Okamura, Rånby, Ito (Eds.): Macromolecular Concept and Strategy
© Springer-Verlag Berlin · Heidelberg 1996

wisdom, various sets of knowledge are strictly and thoroughly screened by sets of experience, which are stored in our brains, and then the selected sets of knowledge are extended to apply to a subject-matter of applied science and technology.

It then follows that the quality and the quantity of wisdom actually depend on those of knowledge and experience, and this dependence may be understood on the analogy of the conception of science or of the paradigm proposed by Thomas Kuhn in 1962 [11.2]. As discussed in the preceding chapters in Parts One and Two, the macromolecular concept or the „paradigm" of macromolecular science has been realized and, eventually, well accepted in this century.

Knowledge is, naturally, based on the five senses of the body, and wisdom needs to be gained through several modes of human action (perhaps five or so modes like making, writing, speaking, seeing, and hearing, where the latter two overlap with the five senses). Despite their mutual difference, however, knowledge and wisdom may be connected with each other or fused together, as two pieces of thin paper are adhered together.

Knowledge in a static behavior is connected to the „concept" in consciousness, but wisdom in a dynamic behavior, or movement, should be connected to a „strategy" in an action. The preceding chapters in Part Three present some discussions of the macromolecular strategy, or, strictly speaking, some details of the tendencies of strategies in the macromolecular disciplines in materials and chemical industry [11.3].

11.3
Importance of the Fusion of Concepts

11.3.1
Concept and Theory

Usually, „concept" is not identical with „theory". Concept is an expression wider and more rough in meaning than theory. For instance, in his second book published in 1977, Kuhn re-defined „paradym" as a concept but not as a theory; his original expression for „paradigm" was rather wrong in his first book in 1971. In short, a concept in one discipline is roughly applicable for another, whereas theory is strictly for a particular discipline usually.

11.3.2
Strategy and Tactics

Usually, „strategy" is not identical with „tactics"; namely, strategy is for the whole, whereas tactics is for parts. Relative to our discussion herein, strategy is for multiple disciplines or international, whereas tactics is for a particular discipline alone or rather domestic (e. g., an international strategy vs domestic tactics; a strategy of physical sciences vs a tactic of chemistry). In this sense, the relation between strategy and tactics is similar to that between concept and theory.

11.3.3
Fusions

Despite the difference in definition, fusions may be possible between concept and theory, strategy and tactics, and concept and strategy, because the two in each pair share common elements. In addition, concept and strategy have a common character in that a concept and a strategy developed for one discipline are useful for another in introducing, respectively, more knowledge (as a static element) and more wisdom (as a dynamic element).

11.3.4
Knowledge vs Wisdom, and Science vs Arts

As discussed above, knowledge is an element of wisdom; namely, wisdom consists of knowledge and experience. Similarly, science is essentially individualistic in process, whereas arts are totalitarian. As Kuhn pointed out in his second book, science and arts are not the same in goal but similar in process. In addition, a process and a goal are similar in arts but different in science. We have discussed the relation between science and arts more in detail in Chap. 4.

11.4
Physical Science as Body and Psychological Science as Mind

From the viewpoint of the body-mind problem discussed above [11.4] all of the traditional disciplines of physical science, including physics, chemistry, biology, and medicine, have been developed in body-oriented directions. Recently, however, new trends have been appearing in some fields of science such as physics (new physics or pseudo-physics), psychology (transpersonal psychology), and medicine (psycho-somatic medicine). In my view, a majority of the contemporary physical scientists are still inclined to feel a risk or a hazard in applying psychological ways of thinking to their own disciplines, mainly because of the still persisting mixed or misled under-standing of psychics (i. e., psychics is entirely different from psychology, but they are often confused).

For the time being, therefore, I would think that, for scientists, a weak separation between physical sciences (body-oriented) and psychosomatic science (mind-oriented) is better than a strong adhesion. Incidentally, I think that „psycho-somatic" is better than „psycho-" to express an intimate „fusion" of body and mind, as shown in the last part of Chap. 10.

11.5
Human Spirit Through Macromolecular Conception

In the classic literature, there are famous phrases of Plato: „The fire of Prometheus should be combined with the spirit of Zeus." In a crude analogy, the former might be taken as technical knowledge and the latter as human wisdom. The „fire" could be obtained by our highly enthusiastic macromolecular investigations, whereas the „spirit" by our highly potential macromolecular conception, and both have already been accumulated by our efforts in these several decades.

In Chaps. 1, 4 and 11, I have explicitly described and emphasized the intimate fusion of macromolecular conception with humanity, namely the intimate „adhesion" between the „front-side", or functionality, of polymeric materials and the „back-side", or sensitivity, of polymeric living body [11.5]. In some years, however, part of my views might be revised due to progresses in brain science (for the body problems) and in psychological sciences (for the mind problems). The principle of the fusion between body and mind should ever be, I believe, comparable to the fusion of macromolecular concept and strategy.

To conclude this book, we, the six coauthors, sincerely hope that readers will understand our macromolecular concept and strategies that involve various „fusions" (see also Chap. 10). We would also be grateful to readers if they would kindly inform us of new propositions on the subjects discussed in this book.

11.6
References

11.1 See also the introduction of the paper in Mountcastle VB (1991) Proc Am Philosophical Soc 135 (4) : 510, Okamura S (1991) K K 10 (9) : 430 (K Ei No. 124)
11.2 Kuhn TS (1970) The structure of scientific revolution. Univ Chicago; Kuhn TS (1977) The essential tension. Univ Chicago, Chap. 12
11.3 Many scientist might consider that both conceptions (in static and strategy in active meaning) are common in the polymeric paradigm.
11.4 Bunge M (1991) Proc Am Philosophical Soc 135 (4) : 513
11.5 Okamura S (1991), Kagaku ni asobu (Enjoy Science), PHP Res pp 73 and 182

Index